INDOOR BEEF PRODUCTION

RON HARDY
and
SAM MEADOWCROFT

FARMING PRESS LIMITED
Wharfedale Road, Ipswich, Suffolk, IP1 4LG
United Kingdom

First Published 1986

British Library Cataloguing in Publication Data

Hardy, Ron
   Indoor beef production.
   1. Beef cattle
   I. Title   II. Meadowcroft, Sam
   636.2'13   SF207

ISBN 0–85236–160–2

Phototypeset by Input Typesetting Ltd, London
Printed in Great Britain by
The Camelot Press Limited, Southampton

INDOOR BEEF PRODUCTION

Contents

Preface

FOR TOO long many farmers have regarded beef production as a secondary enterprise unworthy of adequate inputs of thought or management. For too long moderate animal performance and unattractive financial returns have epitomised this depressed sector.

The development of the barley beef system in the 1950s was a portent of things to come. It was followed by several other planned systems, primarily involving the finishing of housed cattle on rations based on high-quality home-grown forages such as maize and grass silages.

The importance of these developments is three-fold. First they give the advantages inherent in planned systems such as the setting and achievement of input and output targets. Secondly the indoor systems allow greater control of the feeding and protect the animals from the vagaries of the weather, for the first time permitting a high degree of precision in beef production. Thirdly they allow beef producers to take full advantage of the natural superiority of the bull, which efficiently converts feed into beef without the use of artificial growth promoters.

This book is for all those interested in the application of modern techniques to beef cattle husbandry. It shows the beef producer how to rival the dairyman in the efficient use of land—using methods based on sound husbandry and the best stockmanship.

Acknowledgements

THIS BOOK was conceived at Rosemaund Experimental Husbandry Farm (EHF), Hereford, where the authors worked together. We have received much help from Rosemaund colleagues, in particular from Harry Grundy who criticised our drafts.

We wish to thank David Wilks of the Health and Safety Executive and many ADAS (Agricultural Development and Advisory Service) colleagues for checking the manuscript. These include Alan Adamson (Nutrition Chemistry), Bill Hall (Farm Management), Tony Lee (Land and Water Service), Eddie Winkler (Veterinary Investigation Officer), Joe Johnson (Agronomy) and Brian Scott (Livestock Husbandry).

Above all we are deeply indebted to Geoff Barnes (Regional Livestock Husbandry Adviser, Wolverhampton) for his numerous helpful suggestions.

Thanks also to typists Mary Munden, Janet Owen and Maisie Anniss who undertook the original typing.

The authors have drawn heavily on Ministry of Agriculture, Fisheries and Food (MAFF) publications in the writing of this book and wish to acknowledge support received from the Ministry.

Lastly special thanks are due to our wives Pam Hardy and Pat Meadowcroft for assistance in checking manuscripts and for their forbearance over the last few months.

RON HARDY

SAM MEADOWCROFT

April 1986

Chapter 1

WHY BEEF INDOORS?

Indoor Advantages

THE EXPERIMENTAL EVIDENCE indicates that where grassland receives more than 200 kilogrammes of Nitrogen per hectare (kg N/ha), a cutting management will enable more young beef cattle to be kept than under a grazing regime and will maintain or increase individual animal performance. It will also improve the botanical composition of the sward.

Many feeds may be more conveniently fed indoors because this reduces the level of feed wastage. This is true of cereals and also of root crops such as fodder beet. The feeding of roots indoors (which can now be mechanised) brings important advantages in protecting the animals from the vagaries of the weather and protecting the land from poaching.

In Table 1.1 four indoor beef systems are compared with one grazing system (eighteen months beef) in terms of land use and carcass production per hectare.

Table 1.1   Average carcass output of five beef systems

	Cereal beef	Maize beef	Grass silage beef	Fodder beet beef	18 months grazing beef
Area per beast (ha)					
Forage	—	0.17	0.13	0.08	0.31
Cereals	0.31	0.13	0.12	0.12	0.15
Total	0.31	0.30	0.25	0.20	0.46
Mean carcass weight (kg)	238	275	243	240	270
Carcass weight/ha of forage and cereals (kg)	768	917	972	1200	587

NB: Figures based on the period three months to sale.

Grassland Potential

The theoretical potential yield in the best British conditions is 29 tonnes (t) of grass dry matter (DM)/ha/annum. In practice grass yields

11

of 20 t DM/ha/annum ('D' value 65) have been obtained by cutting heavily fertilised Italian ryegrass swards. It is estimated, however, that in our lowlands the average grass yield is 5–7t of DM/ha/annum. The upper figure is reached on dairy farms and the lower figure on farms where grass is utilised by beef cattle and sheep. This means that average grass yields realise only about 30 per cent of grass growth potential.

This large discrepancy between potential and actual production is found also in the utilisation of grass. In grazing trials carried out over four years on six ADAS Experimental Husbandry Farms (EHFs) in the 1970s utilisation varied between 54 per cent and 93 per cent of potential. Data recorded by the Milk Marketing Board (MMB), the Meat and Livestock Commission (MLC), and the Grassland Research Institute (GRI) indicated around 50 per cent utilisation, and 60–70 per cent may be safely considered above average.

The Importance of Grazing in the United Kingdom

During the grass growing season grazing is the cheapest way to feed beef cattle. The relative importance of grazing and conservation is shown in Table 1.2 which indicates that 67 per cent of our lowland grass area is utilised solely by grazing.

Table 1.2   Area under grass in the United Kingdom in 1984

	Million ha
Grass under 5 years old	1.79
Grass 5 years old and older	5.11
Total (excludes rough grazing)	6.90
Area grazed and never cut	4.60 (67%)
Area both cut and grazed	2.30 (33%)

Source: MAFF.

The Case against Grazing

Grazing heavy growths of grass by young beef cattle is notoriously inefficient. Irish trials have shown that, under what was considered to be good grazing management, only 40–60 per cent of the herbage was eaten, the remainder eventually rotting and returning to the soil. Under average conditions 25–30 per cent wastage of the growth of pastures may be expected. The main causes seem to be:

- *The restless behaviour* of the animals at turnout to grass. This contributes to the check to liveweight gain which characterises the first four to six weeks of the grazing period.

- *Treading damage.* Reductions of 20 per cent in herbage growth have been reported even where no damage was observed. This figure can be 30 per cent or more where grassland is densely stocked in wet conditions.

- *Fouling* by urine and dung results in areas of grass being rejected by young beef cattle. Such areas have totalled 40 per cent of the whole in late season, where cattle were paddock-grazed at Rosemaund EHF. Urine scorch can also cause long-term pasture damage.

- *Declining quality* of grassland in late season means poorer nutritional value and lower intakes. This occurs as over-mature material accumulates.

- *Worm burden.* Most pastures harbour large numbers of parasitic worm larvae which limit the performance of grazing cattle. The only exceptions are new leys and swards not grazed by cattle in the previous twelve months.

- *Variable forage intakes.* During periods of bad weather dry matter intakes are reduced.

The target liveweight gains set for young cattle at grass are around 0.8 kg/day over the whole season. This level of performance is, however, not always achieved, requiring as it does the availability throughout the grazing season of palatable forage of high digestibility. Typical gains achieved at grass by young cattle are shown in Table 1.3 which is based on performance in sixty-four units during 1984. The nitrogenous fertiliser applied was at the rate of around 200 kg N/ha.

Table 1.3 **Grazing performance and stocking rates (eighteen months beef)**

	Friesian steers	Hereford x Friesian steers	Hereford x Friesian heifers
Daily gain at grass (*kg*)	0.71	0.70	0.60
Grazing stocking rate (*kg/ha*)	1579	1332	1285
Grazing gain (*kg/ha*)	736	649	570
Concentrates (*kg/ha*)	83	92	50

Source: MLC Beef Improvement Services. Data sheet 85/2, 'Beef cattle at grass'.

This modest performance by young beef cattle at grass was not the consequence of heavy stocking rates. There was no evidence that cattle performance was better in units with the lower stocking rates.

All too frequently the progress of grazing cattle is too slow to achieve the target weights at yarding even where supplementary concentrates are fed at grass to boost inferior performance.

Many attempts have been made to devise improved grazing systems which would secure good liveweight gains combined with high stocking rates. Perhaps the most important have been rotational paddock grazing, the integration of conservation and grazing as in the $^2/_3$–$^1/_3$ system, the 'follow N' system and leader/follower grazing. All of these allow rather more control over grazing management than does continuous stocking. Unfortunately, these systems in most cases have not significantly improved the performance of young beef cattle at grass. We are led to the conclusion that on the whole attempts to improve the grazing management of young cattle have been unsuccessful. Grass conservation perhaps has more to offer.

The Case for Grass Conservation

Conservation is an essential part of intensive grassland management and achieves the following objectives:

- The exploitation of the high potential of short-term Italian ryegrass cutting leys.
- The effective utilisation of the spring flush of grass growth.
- Ensuring the supply of worm-free aftermath grazing when grass keep is in short supply in mid-season.
- The provision of high quality bulk forage for winter feeding.

Comparison of Output under Grazing or Cutting Management

There have been few valid comparisons, but fortunately there is valuable evidence from a series of Agricultural Development and Advisory Service (ADAS)/GRI trials, and from the North of Scotland College of Agriculture.

ADAS/GRI trials
Carried out on six EHFs over a four-year period these experiments compared the response of a perennial ryegrass sward to fertiliser N at input rates of 200, 400, and 600 kg N/ha. The comparisons were made under cutting-only or intensive paddock grazing-only management systems.

The response to nitrogen applications was always greater under cutting than grazing. At the lowest level of N the DM yields from grazed plots were higher than those from cut plots in only eight of twenty-four comparisons. At higher levels of N grazed plots always yielded less than cut plots. There was a tendency after four years for the proportion of sown species in the grazed swards to be less than that in the cut swards.

North of Scotland College trials

These compared various summer managements in terms of animal output. In 1980 and in 1981 the storage feeding of beef cattle was compared with field grazing and with zero grazing. Zero grazing is an American term which means that forage is cut and carted to housed stock throughout the grass growing season. Storage feeding is defined as a non-grazing system with all-year-round housing and feeding. Although storage feeding did not necessarily increase individual liveweight gains it did allow a heavier stocking rate.

In the 1982 trial Friesian bulls were allocated to each of the following treatments:

Storage fed silage plus barley supplement.
Storage fed silage plus barley/fishmeal supplement.
Zero grazed.
Field grazed.

Differences in daily liveweight gain were again small. However, there was a considerable variation in the calculated stocking rates, these being approximately seven, nine and twelve animals per forage hectare for field grazing, zero grazing and storage feeding respectively.

In the 1983 trial Friesian steers were used and storage feeding and different grazing managements were compared. Two grazing groups had their feeding buffered by the availability of extra grass or extra silage (see Table 1.4).

Table 1.4 Performance on four summer managements

	Storage-fed silage	Grazed	Grazed + grass buffer	Grazed + silage buffer
Liveweight gain early summer 2 May–3 July (kg/day)	1.21	0.90	1.32	1.20
Liveweight gain late summer 3 July–19 Sept (kg/day)	1.42	1.23	1.23	1.32

Source: North of Scotland College of Agriculture

In this trial storage feeding gave better liveweight gains than grazing throughout the season, and better gains than buffered grazing from mid season onwards. The advantage to storage feeding over grazing was 34 kg of liveweight gain over the season.

Zero grazing and storage feeding are two techniques which may be used to achieve the advantages of indoor forage feeding noted at the beginning of this chapter.

Zero Grazing

The advantages of zero grazing are:

High quality forage may be fed daily with the avoidance of the nutrient losses inherent in silage-making.

The sward damage and fouling caused by grazing are avoided.

The cattle are not exposed to the parasitic worm burdens of grazed swards.

Stocking rates may be increased by 25–35 per cent while maintaining individual liveweight gains.

Unfortunately, zero grazing incurs high labour and machinery costs, as daily harvesting is required. It is most easily justified on fragmented farms or on land which is readily poached. In theory it offers the potential for extremely efficient grassland utilisation during the spring and summer months. The biggest snag is the high level of expertise required to operate a zero grazing system in such a way that an adequate supply of young grass is available every day. This is beyond most mortals whereas storage feeding presents no comparable problems.

Storage Feeding

Storage feeding gives all the advantages listed for zero grazing except that the losses involved in the ensilage process must be borne. Unlike zero grazing it operates the year round and is very much simpler. This is because with storage feeding the important decisions are taken only every four to six weeks at the time of the ensilage cuts, whereas zero grazing requires daily decisions throughout the grass growing season. Two storage feeding systems are described in Chapters 6 and 7.

Summary

Housed systems invariably out-perform grazing systems in their ability to produce carcass meat per hectare of land devoted to the enterprise. It can be seen from Table 1.1 that compared with eighteen months beef the advantage is 31 per cent for cereal beef, around 60 per cent for the silage feeding systems, and approximately 100 per cent for fodder beet beef.

   The conclusion must be that current grazing systems do not achieve good feed utilisation. Food wastage can be reduced and stocking rates and carcass production thereby increased by keeping cattle indoors using methods described in this book.

Chapter 2

THE PERFORMANCE AND BEHAVIOUR OF BULLS KEPT INDOORS

A VERY IMPORTANT decision to be made by operators of intensive beef systems is whether to purchase bull or heifer calves. Those who choose bulls must then decide for or against castration.

Historically, most male calves have been castrated. Dr Joe Harte of the Grange Research Centre, Southern Ireland, and others have given the following reasons for this:

Castrates were more docile for draft purposes.

Hard manual work required a high fat diet.

In the past animals were taken to a greater age at slaughter and steers were more controllable than bulls.

Castration prevented indiscriminate breeding.

Steers finish more readily than bulls.

It is interesting that the first four reasons have no relevance in a modern indoor beef unit and the fifth is not so important in intensive units where the diet is closely controlled. The castration operation may cause a temporary setback in performance, but more important are the long-term effects of removing a source of natural growth promoting hormones.

On the Continent large numbers of entire cattle are now finished. Bull beef production in the United Kingdom is expanding and ADAS estimates that bulls amount to 6–7 per cent of all beasts slaughtered in late 1985. This increase is a result of the recent development of a number of indoor beef production systems and also because of increased consumer demand for lean meat. Bull beef producers gain substantial benefits of improved growth rate and Food Conversion Efficiency (FCE) but must contend with the problems involved in the safe handling of bulls and in the marketing of the meat.

The Comparative Performance of Bulls and Steers

The superior performance of the bull as a beef animal is well documented. In terms of daily liveweight gain and FCE the performance of the entire animal, in intensive units, is frequently some 10 per cent better than that of the steer. Bulls are later maturing than steers and can be taken to heavier weights and still produce lean carcasses with high saleable meat yields.

In 1962 J. D. Turton reviewed the evidence to date. He concluded that a variety of breeds of bulls gained 0.05–0.36 kg/day more than comparable steers during the finishing period. The bulls were also more efficient converters of food to meat than the steers.

The most thorough series of comparisons in the United Kingdom was carried out by the Agricultural Departments and Advisory Services and reported by Wickens and Ball in 1967. This work, at forty-six centres in England, Wales, Scotland and Northern Ireland, compared the behaviour, productivity and carcass quality of groups of yard-fed bulls and steers. The comparative performance of those kept indoors throughout their lives is shown in Table 2.1.

Table 2.1   Performance of bulls and steers on indoor systems

	Steers	Bulls	% advantage (Bulls over steers)
Daily liveweight gain (kg)	1.13	1.25	10.6
Days on test	254	236	7.1
Weight at sale (kg)	405	414	2.2
Carcass weight (kg)	227	233	2.2
Killing out %	56.0	56.3	Similar
Compound intake (kg)	1669	1530	8.3
Feed Conversion Efficiency (kg compound per kg LWG)	5.87	5.20	11.4

LWG = Liveweight gain
Source: *Experimental Husbandry 1967*, No. 15, 64–87.

The housed bulls showed an 11 per cent higher rate of liveweight gain and FCE than comparable steers. The better feed conversion of the bulls resulted in a saving of 139 kg of compound per head and they were sold eighteen days earlier than the steers even though they were slightly heavier at slaughter.

The considerable advantages in performance consistently achieved by bulls kept indoors were not matched by bulls on grazing systems. Wickens and Ball reported that Friesian bulls which spent one summer at grass showed only half the advantage over steers achieved by bulls kept indoors. At Rosemaund EHF groups of Friesian and Hereford *x* Friesian bulls and steers, purchased in 1970 and 1971, were compared on the eighteen months beef production system, which included one grazing season. Because of their unsettled behaviour at grass the bulls were no heavier than the steers when yarded as yearlings, and they subsequently required high compound inputs to achieve a satisfactory finish.

Keeping bulls on grazing systems is a dubious practice involving practical and safety problems, yet offering negligible advantages over the steer. However on housed systems the bull may be safely contained and moreover is clearly the better performer since it is suited to the indoor environment, and its full potential is realised on the high energy diets fed in housed units. In this situation castration may reduce the financial margin per head by about 20 per cent. This results from higher feed costs caused by poorer FCE and a lower saleweight.

Marketing
Finished steers can be sold through livestock auctions or at deadweight centres whereas the market outlets for bulls are more limited. The majority are sold 'on the hook' but sales through livestock markets are increasing.

There is a long standing resistance to bull beef in some sections of the meat trade. There were sound reasons for this in the past including lack of finish in some bull carcasses and the possible occurrence of 'dark cutting meat'. However we now know that good management can prevent these problems and much of the opposition to young bull-beef appears to be based on ignorance and prejudice.

Carcass quality of bulls and steers
Many comparisons have been made of the carcasses and of the meat quality of bulls and steers and they are remarkably similar. For instance the killing-out percentage of bulls and steers has been almost the same in most comparative trials. In the majority of these the bulls have shown a slightly higher figure but the advantage either way has rarely exceeded 1.5 per cent.

As bulls mature forequarter development increases markedly. However, in housed cattle slaughtered young, the proportion of additional forequarter in the whole carcass is often less than 1 per cent. For instance, Wickens and Ball reported that the bulls' forequarters constituted 50.2 per cent of the carcass compared with 49.8 per cent in the steer.

The proportion of lean, fat and bone in the carcass are of prime importance to the butcher and housewife. The figures in Table 2.2 are taken from Meat Research Institute (MRI) results using bull/steer twins slaughtered at 400 days. They were reported by A. V. Fisher and others in 1986, and others have recorded similar results.

Table 2.2 Percentages of lean, fat and bone in carcasses

	Bulls	Steers
% lean	65.7	60.1
% fat	17.8	23.8
% bone	16.5	16.1

Source: MRI

These results show that bull carcasses contain much more lean and much less fat. Since fat is expensive to produce in terms of feed inputs and is not wanted by most consumers the above figures are a strong recommendation for bull beef.

The higher proportion of the forequarters does mean that the proportion of high-value cuts in the bull carcass is slightly lower but this is more than balanced by the higher proportion of saleable meat from the bull carcass.

Eating quality of bulls and steers
The comparative eating quality of bull and steer meat is a contentious subject. Some buyers have had little experience in handling meat from young bulls. Their contention that bull beef is inferior is based on their experience of the low-quality beef from breeding bulls which are culled at an advanced age.

Much of the evidence on meat quality has come from comparative assessments by trained taste panels at the MRI, Bristol.

The taste panels have occasionally marked bull beef a little lower than steer beef for tenderness, but they have detected no difference in flavour or juiciness. When the MRI conducted a survey involving over 150 households they found that consumers could detect no significant differences between bull and steer beef. Neither taste panels nor consumers have reported any taint in bull beef.

Bulls lay down fat later than steers, therefore their carcasses are frequently less well finished. Thus bull carcasses appear somewhat darker. Lack of finish can be prevented by ensuring that the energy intake is adequate to allow bulls to express their full potential. This is achieved by increasing the level of supplementary feeding to bulls on forage based diets in the last few weeks before slaughter.

The effect of diet on human health has received much publicity, with the most attention being given to recommended levels of fat intake in relation to coronary heart disease. The contribution of energy from fat as the proportion of total calories is around 40 per cent in the United Kingdom. In 1984 the Committee on the Medical Aspect of food policy (COMA) advised on the relationship between diet and cardio-vascular disease and recommended target intakes of 35 per cent of food energy from total fats and 15 per cent of food energy from saturated fat.

The COMA panel also recommended that 'consideration should be given to ways and means of encouraging the production of leaner carcasses in sheep, cattle and pigs.' This advice reinforces the present consumer preference for lean meat and should increase the saleability of lean bull beef.

The dark cutting meat problem

Bulls are more prone than steers or heifers to produce dark cutting carcasses as opposed to the pink colour favoured by housewives. Dark meat has a relatively short shelf life and is less acceptable to consumers and must be sold at a lower price. It is now known that this problem is not intrinsic to the bull, but is caused by pre-slaughter stress. Dark cutting can be eliminated by the quiet handling of bulls on the farm, in transit and at the abattoir, with no mixing of groups of cattle in any of these three situations. Immediate slaughter on arrival at the abattoir is much preferable to an over-night stay in the lairage.

There is no doubt that intensively reared bulls finished at under fifteen months can provide lean beef which is completely acceptable to the consumer. It is valued for vacuum-packed joints and suitable for the export trade to the Continent.

Bull Behaviour

Bulls are naturally more individualistic, restless and aggressive than steers. The grazing situation gives much scope for this restlessness, but experience has shown that bulls kept indoors are much calmer and easier to manage. The high liveweight gains associated with housed systems ensure that bulls are sold before the worst behavioural traits develop.

Wickens and Ball reported that bull behaviour was satisfactory on thirty-one of forty-six farms observed. Riding and aggression were noted in a minority of these units. Harte found little problem during ten years experience in Ireland. Most producers report that riding is normal in groups of bulls, but that dangerous aggression is mostly limited to Friesian and Holstein bulls from ten months of age onwards. This has been the experience at Rosemaund EHF where around one per cent of the Friesians handled have been the only animals to jeopardise staff safety.

The availability of suitable housing (preferably in a quiet situation) and handling facilities is vital (see Chapter 3). Above all, sensible management practices will minimise the problems which bulls can pose.

Group size is important; the smaller the better. In practice groups of twenty or less have been found to be satisfactory. The appropriate stocking density indoors for bulls is the same as that for steers of the same size, but in bedded houses bulls may require an extra 20 per cent straw compared with heifers or steers. Normally the bulls in a pen should be matched in age and size as recommended in the Welfare Code for cattle. It is clear that the mixing of groups or the reintroduction of bulls after a period of absence will lead to restless behaviour, as will the proximity of female cattle, particularly mature cows.

All bedding and handling operations can be guaranteed to excite bulls, and any disturbance should therefore be kept to a minimum. Staff tending bulls should be consistently vigilant, quiet and firm.

The Safe Handling of Bulls

Everyone working in agriculture has far-reaching responsibilities under the *Health and Safety at Work etc. Act 1974*. This places general obligations on employers to ensure so far as is reasonably practicable the health, safety and welfare at work of all their employees. In addition, employees and the self-employed have a duty in respect of their own health and safety and that of other people. It is essential therefore that safe buildings, fixtures and handling facilities are provided and that all concerned adopt safe working practices.

Anyone inexperienced in bull beef production should seek prior guidance on health and safety matters from HM Agricultural Inspectorate at the local office of the Health and Safety Executive. The following publications (available from the Health and Safety Executive) will help bull beef producers to fulfil their obligations under the above Act and ensure the safety of all:

Health and Safety Executive Guidance Note G S 35: 'Safe Custody and Handling of Bulls on Farms and similar Premises'.
Health and Safety Executive Leaflet A S 3: 'Agricultural Safety – Bulls'.

Experience has shown that given well-designed buildings and facilities, sensible management routines and experienced, cautious staff, there are few safety problems in the housing of bull-beef cattle.

The Control of Bull Behaviour

This may normally be achieved in housed bulls by the management methods outlined above. There are in addition four techniques (short of full castration) for the treatment of bulls with the objective of combining the best features of bulls and steers but all may have welfare implications. Three of these are Russian castration, immuno-logical castration and induced cryptorchidism. The fourth is the use of hormone implants, and because of the possible ban on these, their use on bulls, steers and heifers is described in the Appendix.

Russian castration

The Russian method of castration involves the removal of only the sperm-producing part of the testicle, leaving the outer hormone-producing part intact. The objective is to combine the high growth rate and food-saving attributes of the bull with the calmer behaviour of the steer.

Two of the EHFs taking part in the trials reported by Wickens and Ball included groups of Russian castrates, and Table 2.3 details their performance and that of comparable steers and bulls.

Table 2.3   Performance of steers, bulls and Russian castrates

	Steers	Bulls	Russian castrates
LWG/day weaning to slaughter (kg)	1.08	1.23	1.18
Days—weaning to slaughter	306	270	282
Feed consumption, weaning to slaughter:			
Barley mix (kg)	1888	1694	1711
Hay (kg DM)	129	107	118
Carcass weight (kg)	229	225	226

Source: *Experimental Husbandry*, No. 15, 64–87.

Russian castrates showed an improvement in growth rate of over 9 per cent compared with steers; but their performance was 4 per cent less than bulls. Total feed intake of the Russian castrates was similar to that of the bulls, and 177 kg less than that of steers to produce carcasses of similar weights. The Russian castrates differed from bulls in that there was an absence of riding or aggressiveness.

On available evidence it is unlikely that the saving in feed cost achieved by Russian castration will be sufficient to cover the cost of the operation.

Immunological castration

It is known that the secretion of testosterone can be controlled by immunising bulls against a specific hormone. This results in low serum testosterone levels, reduced sexual activity, reduced semen production and more docile behaviour. The effects on testosterone production and behaviour lasted for four to six months at the Rowett Research Institute. However the animals were found to be infertile for fifty weeks after treatment and at slaughter a few sperm were found but these were dead. This technique is still under development, but may gain a commercial sponsor in the near future. Its effect is to cause bulls to behave more like steers for a limited period of time with little effect on performance. It may have a place in calming the behaviour of bulls kept at grass for one season on the eighteen months beef production system, but more experience is needed before its wider use can be predicted.

Induced cryptorchidism
This is achieved by pushing the testes to the top of the scrotum where they are retained by placing an elastrator ring immediately below them. Published results on the effects of this technique have in general been confirmed by E. D. Rees and I.A.M. Lucas who used forty-eight beef cattle to compare bulls, induced cryptorchids and steers at the University College of North Wales, Bangor in 1973–76. In live-weight gain and age at slaughter Friesian cryptorchids tended to perform better than Friesian steers but poorer than Friesian bulls. Hereford x Friesian cryptorchids performed in a similar manner to bulls of this cross. The cryptorchids were as restless and aggressive as the bulls, although they were almost completely sterile. They resembled the bulls in most carcass characteristics.

The technique has theoretical attractions in suckler herds in reducing the risk of unwanted pregnancies. However the nature of the operation has prevented any significant uptake in this situation.

In intensive units where bulls and heifers are kept apart there would appear to be nothing gained from induced cryptorchidism.

Chapter 3

BEEF CATTLE HOUSING REQUIREMENTS

ESSENTIAL CONSIDERATIONS should include ventilation and animal environment, stocking densities, feeding and drinking facilities, building design, cattle handling facilities.

Ventilation

Respiratory infection is undoubtedly the most common and financially the most damaging disease in housed calves and cattle. It appears most frequently in young stock between ten and twenty weeks of age. This critical period normally occurs when the disease resistance derived from colostrum has diminished and animals have not yet acquired sufficient natural immunity. High levels of mortality can result. In addition, animals which have been infected, and apparently successfully treated, may exhibit subsequent poor performance and even succumb to further disease. This results from chronic lung damage and can cause further losses and more expensive veterinary treatment.

The aim of good building design should be to give protection from extremes of rain or snow and protection from draughts. The design should also ensure frequent air changes. In adequately ventilated buildings harmful pneumonic agents do not build up, noxious gases and dust are dissipated, and wet bedding and dripping roofs are avoided. Only very young calves are likely to suffer from cold stress. Older, well-fed, housed cattle are not susceptible because they produce adequate amounts of metabolic heat. For the lower critical temperatures for various age groups of cattle see Table 4.1.

The livestock house can be ventilated in three ways:

First and most commonly, by wind effect where external air movements across a structure cause atmospheric pressure differences between the windward and the leeward sides. Where adequate openings are provided this will enable 'fresh' air to be brought into the building, in turn displacing the stale or contaminated air. This is generally known as 'natural' ventilation.

Secondly, artificial (or mechanical) ventilation can be used. This is necessary where buildings cannot ventilate naturally. It is, however, seldom applied to buildings housing mature beef cattle.

Thirdly, a system based on the 'stack effect' can be employed. This effect works as follows: air which is heated by its proximity to housed animals expands and is then less dense than the surrounding colder air. It therefore moves upwards and leaves the building at high level. This movement encourages the entry of cooler air into the building at low level. Consequently, two aperture types are required in livestock buildings, air inlets and outlets, with a height difference between them. This is shown in a simplified manner in Figure 3.1.

It should be stressed that the most satisfactory results are obtained when the building design allows benefits of both 'natural' and 'stack effect' ventilation.

The outlet
This is most important in controlling natural ventilation. In modern livestock housing, using a dual pitched roof, outlet ventilation can be provided satisfactorily by an open 'Venturi' ridge (Figure 3.2). The importance of this is often underestimated. This figure describes a 'Venturi' ridge which performs well for both stack and wind-effect ventilation. Other profiles, even though providing the same open area, may perform badly under certain conditions. Particular wind directions, for example, can transform the ridge outlet to act as an inlet, introducing down-draughts and forcing rain and snow into the building. The 'Venturi' ridge has been tested in the laboratory and in many practical situations and is strongly recommended. A protected

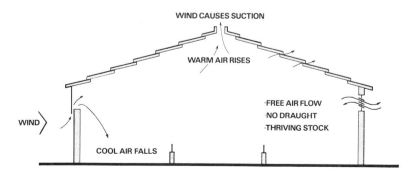

FIG 3.1 : SIMPLIFIED DIAGRAM OF NATURAL VENTILATION
Source: Scottish Farm Building Investigation Unit

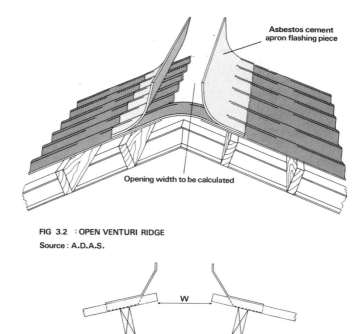

FIG 3.2 : OPEN VENTURI RIDGE
Source : A.D.A.S.

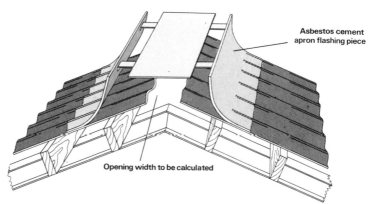

FIG 3.3 : PROTECTED OPEN RIDGE
Source : A.D.A.S.

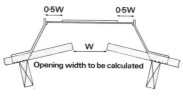

open ridge (Figure 3.3) is also recommended. Whilst performing similarly to the open ridge it gives additional protection from rain and
snow entering feed passages and bedded areas.

The 'breathing' roof and slotted roof (Figure 3.4) are also suitable
designs for achieving efficient outlet ventilation, particularly in wide-
span buildings. They have been known for many years but surprisingly
have been out of favour in recent times.

The inlet
Normally, inlet ventilation is provided below the eaves. Suitable
arrangements include continuous gap, space boarding, perforated
sheeting or honeycombed walling (Figure 3.5). Usual practice
demands that the total area of inlet ventilation is twice that of the

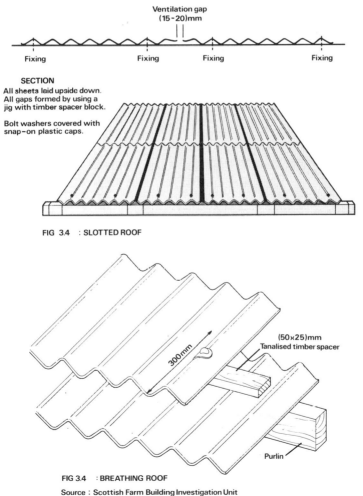

FIG 3.4   : SLOTTED ROOF

FIG 3.4   : BREATHING ROOF

Source : Scottish Farm Building Investigation Unit

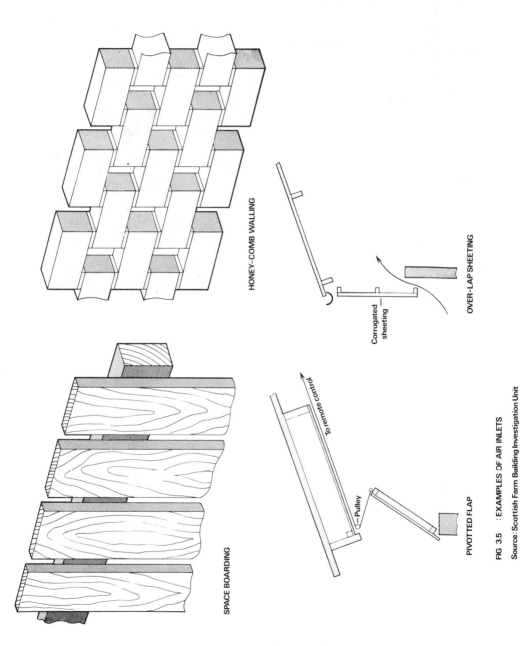

HONEY-COMB WALLING

SPACE BOARDING

OVER-LAP SHEETING

Corrugated sheeting

PIVOTTED FLAP

To remote control

Pulley

FIG 3.5   : EXAMPLES OF AIR INLETS

Source : Scottish Farm Building Investigation Unit

Open ridge outlet and space boarding inlet. Photograph H. F. Grundy

outlet. It is imperative that a balance be obtained between adequate ventilation during still air conditions and the avoidance of draughts (particularly down draughts at stock height) under windy conditions. Inlet positioning and space boarding gap are important. The detailing and calculation of the total areas of inlet and outlet in relation to stocking density are critical to achieve optimum ventilation. Professional advice should be sought before new buildings are erected or old buildings modified. Additionally, once a building has been constructed (or modified) and is fully stocked, the ventilation should be physically checked under various climatic conditions. For example, still air and different wind directions, particularly the direction of the prevailing wind. This will ensure adequate ventilation and the absence of excessive draughts. Environmental monitoring equipment should be used to determine air speeds and directions, temperatures and relative humidities both externally and internally.

Optimal Environmental Conditions

The energy balance of an animal is affected by temperature, air speed, relative humidity, feed intake, wetness of floor and the thickness of the animal's coat. These factors are interdependent. Table 3.1 gives recommendations for temperature, air velocity at stock level and ventilation rate for calves and finishing adult stock.

Table 3.1   Recommendations for temperature, air velocity and ventilation rate

Stock	Temperature Range (°C)	Maximum air velocity at stock level (m/s)	Ventilation per kg liveweight (m³/h)
Calves (all ages)	5–20	0.25	0.75–2.25
Finishing cattle (adult stock)	0–15	0.25	0.5–1.5

Source: British Standards BS 5502.

Reasonable variations outside these limits need not be detrimental to stock welfare. It is not practical to control relative humidity but normally it will remain within acceptable limits if recommended ventilation rates are maintained. Minimum air space requirements for healthy stock of over 200 kg liveweight are given in Table 3.2.

Table 3.2   Minimum air space requirements

Liveweight (kg)	Air space (m³/head)
200	15
300	20
over 300	25

Source: ADAS, *Beef Cattle Housing.*

Gas concentrations should not exceed those given in Table 3.3. Normal ventilation will ensure that they stay below these limits. However, when slurry is removed from below slats, the risk of excessive gas concentrations is greatly increased and maximum ventilation should be provided at this time.

Table 3.3   Maximum allowable gas concentrations

Gas	Concentration for calves ppm (v/v)*	Concentration for adult cattle ppm (v/v)
Carbon dioxide	3000	5000
Ammonia	20	25
Hydrogen sulphide	5	10

* ppm = parts per million
  v/v = volume for volume
Source: BS 5502.

Stocking Rates

These should be calculated on the basis of the animals' projected liveweight at the end of a particular stage of housing.

Bedded yards
Recommendations are given in Table 3.4(a).

Table 3.4(a)   Stocking rates on solid floors (m²/head)

Liveweight	Bedded area* (excluding troughs)	Loafing/feeding area (excluding troughs)	Total area*
200	2.0	1.0	3.0
300	2.4	1.0	3.4
400	2.6	1.2	3.8
500	3.0	1.2	4.2
600	3.4	1.2	4.6

* For fully bedded yards the total area should be used.
Source: BS 5502.

Animals fed on dry feeds (e.g. barley beef) can be kept at stocking rates higher than those given. Where very wet feeds are used (e.g. low dry matter silage or ad lib roots) it is advisable to operate with lower stocking rates than those shown to avoid over frequent bedding-down. Also, it is advisable for cattle of 100–250 kg to be stocked at rates considerably below those shown in Table 3.4(a). In the example shown for follow-on housing (Figure 3.6) total areas of under 4.0 m²/head are shown.

At such stocking rates, the persistent lung problems frequently encountered in this age group have been minimised.

Slatted floors
Recommended stocking rates on fully slatted floors are given in Table 3.4(b), together with minimum slat width and maximum gap width between the slats.

Table 3.4(b)   Recommendations for fully slatted floors

Liveweight (kg)	Stocking rate (m²/beast) (excluding troughs)	Minimum width of slat (mm)	Maximum width of gap (mm)
200	1.1	125	40
300	1.5	125	40
400	1.8	125	40
500	2.1	125	40
600	2.3	125	40

Source: BS 5502.

It is the authors' experience (and that of many farmers) that cattle below 250 kg are not suited to slatted floor housing although in areas of severe bedding straw shortage, young cattle are sometimes housed on slats. However, smaller gaps (30 mm) and widths of slats (100 mm), are advisable. Honeycomb slats are particularly favoured by some producers, who believe that they give better animal comfort than conventional slats.

Obviously slats eliminate straw use. They must be laid level and secured firmly to prevent rocking. Grids of seven slats help to overcome these problems. If they are too smooth, animals may suffer injury due to slipping, if too rough, comfort is impaired. A part-slatted, part-bedded system can be used but it is advisable to use only chopped straw so that gaps do not become blocked.

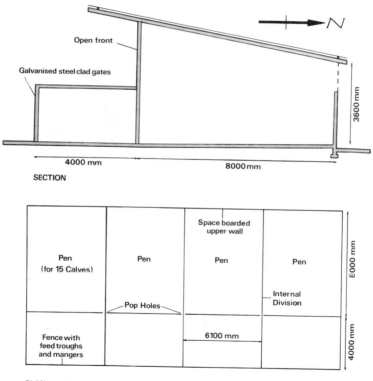

PLAN

FIG 3.6   : FOLLOW ON SHED – FOR YOUNG STOCK

Cubicles

These are now less favoured by beef finishers, particularly where male animals are used. The problem of slurry disposal must be dealt with and overall stocking rates are low. However, cubicles reduce straw usage by up to 90 per cent. Dimensions of cubicles for young stock and older stock are given in Table 3.4(c).

Table 3.4(c) Cubicle dimensions for beef cattle (m)

Liveweight (kg)	Length (including kerb)	Clear width between partitions
75–150	1.2	0.60
150 250	1.5	0.75
250–375	1.7	0.90
over 375	2.1	1.10

Source: BS 5502.

Sloped floors

This is a novel method of housing beef cattle and was developed in the North of Scotland. The lying area, which is fully concreted and has a gradient of one in sixteen, is given only a very light straw littering daily when semi-soiled waste is also removed. A high standard of management is essential and daily labour requirements are high. It is more likely to be suited to the feeding of the drier diets and to the smaller units.

Space allowances are given in Table 3.4(d).

Table 3.4(d) Stocking rates on sloped floors (m²/beast)

Liveweight (kg)	Area (excluding troughs)
200–250	1.75–1.90
300–350	2.00–2.15
400–500	2.35–2.50

Source: Scottish Farm Buildings Unit.

Roofless units

These are best adopted for use in relatively low rainfall areas, that is, below 750 mm. They require a high standard of management and can produce excessive amounts of slurry. For instance 100 mm rainfall on 10 m² floor area produces 1000 litres of water. Floors can be sloping concrete, straw bales over hardcore or slatted. Recommended stocking rates are given in Table 3.4(e).

Table 3.4(e)   Stocking rates for roofless units (m²/beast)

Liveweight (kg)	Straw Yards Feeding	Lying	Sloping concrete	Slats
200	1.0	2.0	1.75	1.1
300	1.0	2.4	2.15	1.5
400	1.2	2.6	2.60	1.8
500	1.2	3.0	2.60	2.1

Source: ADAS, *Beef Cattle Housing.*

Feeding Arrangements

Frontage allowances are related to weight of stock, feed type, and frequency and method of feed delivery. The feeding frontages recommended are given in Table 3.5.

Table 3.5   Feeding frontage (mm/beast)

Liveweight (kg)	Rationed feeding	Ad lib feeding
130–250	300–450	100
250–350	450–550	125
over 350	550–700	150

Source: BS 5502.

Food can be self-fed (e.g. silage) or offered in hoppers, in bunkers, from mangers or in troughs. Some designs of the latter are shown in Figures 3.7–3.9.

In the authors' experience the simple floor design used in Figure 3.9 is to be preferred for the trough base. An arrangement of diagonal steel bars (as in Figure 3.8) fixed over timbers (as in Figure 3.7) is to be preferred for the feed barrier. Any wooden partitions in the troughs should be protected by metal if they are not to be chewed by the stock.

Drinking Arrangements

Water consumption increases in line with animal liveweight. It is further increased by the feeding of high dry matter diets and by high temperatures in the house. Water troughs or bowls should be positioned to avoid feed contamination. They should be adjustable in height in bedded yards and protected by a rump rail fixed 800 mm above bed level to prevent fouling. Recessing them into the wall can also be a solution. The position of the bowl or trough should be easily accessible to the stockman. It is very difficult to prevent spillage and so the drinkers should be positioned to avoid wetting the lying areas.

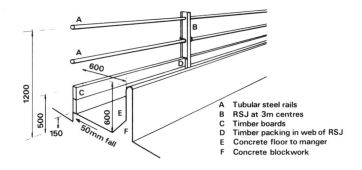

A Tubular steel rails
B RSJ at 3m centres
C Timber boards
D Timber packing in web of RSJ
E Concrete floor to manger
F Concrete blockwork

FIG 3.7 : RAIL BARRIER & COMPLETE DIET BUNKER
Source: A.D.A.S. booklet 2512

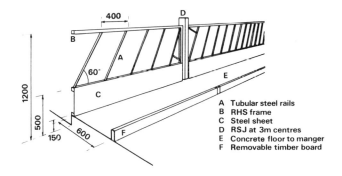

A Tubular steel rails
B RHS frame
C Steel sheet
D RSJ at 3m centres
E Concrete floor to manger
F Removable timber board

FIG 3.8 : DIAGONAL RAIL BARRIER USED AS AN ACCESS GATE
Source: A.D.A.S. booklet 2512

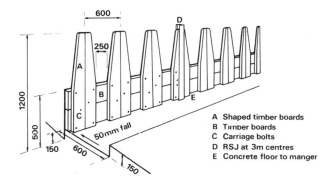

A Shaped timber boards
B Timber boards
C Carriage bolts
D RSJ at 3m centres
E Concrete floor to manger

FIG 3.9 : TOMBSTONE BARRIER
Source: A.D.A.S. booklet 2512

Attention should also be paid to protection against frost by the provision of insulation and drainage stop cocks. Estimates of water consumption by mature stock are given in Table 3.6.

Table 3.6   Water provision

Consumption per head per day (litres)	Number of bowls per 20 beasts	Surface area of trough per 20 beasts (m²)
50	2	0.3

Source: ADAS, *Beef Cattle Housing.*

Animal Waste

The containment and disposal of animal waste must be arranged to avoid the pollution of water courses. Average production is given in Table 3.7.

Table 3.7   Production of waste

Liveweight (kg)	Approximate volume of waste/head (litres/day)* (faeces + urine)
85–140	7
140–330	14
330–450	21
over 450	32

* 1000 litres = 1 m³.
Source: BS 5502.

In slatted units faeces and urine tend to separate in the storage tanks below the slats. Experience gained at Rosemaund EHF has shown the following:

- Slurry agitation prior to removal is essential. A chopper/mixer slurry pump is ideal for this purpose.

- Slurry removal is facilitated by dilution with water.

- Toxic gases (particularly hydrogen sulphide) are emitted during agitation, and cattle should be removed before this takes place.

- The addition of silage effluent to slurry increases the production and emission of poisonous gases. *Fatalities have occurred as a result of entering cellars under slatted houses.*

Underslat scrapers (as used at Liscombe EHF) or tractor fore-loaders working under high level slats are also effective.

Beef Building Design

A suitable design for cattle up to eight months old is given in Figure 3.6 and four basic designs for cattle from six months to slaughter are shown in Figures 3.10–3.13.

It is strongly recommended that animals of widely varying ages should not be housed in the same air space. Therefore indoor operators should house cattle up to six to eight months in rearing buildings (see Figure 3.6, page 34) and then move them to finishing accommodation (Figures 3.10–3.13) from this age onwards. By doing so two important advantages are gained.

1. Cattle at the most vulnerable age for pneumonia do not share a building with older stock. This sharing is a thoroughly undesirable practice which frequently occurs in large span umbrella buildings. It is a recipe for lung problems.

2. Floor space is not wasted in the younger growing stages, so the temptation to overstock in large buildings is avoided.

The four basic examples for older stock are: fully bedded with raised feed passage; part bedded with scraped passage behind the feeding barrier; fully slatted over a storage chamber, and sloped floor.

There are many permutations on these designs involving variations particularly in feeding facilities.

1. *Fully bedded with raised central feeding passage* (Figure 3.10). This type of housing is extremely popular in areas with plentiful straw supplies. Its main disadvantage (other than the high straw usage) is the difficulty of stock movement and access to pens and as muck builds up, the mangers may become contaminated and feed intake may be restricted.

2. *Part bedded/part scraped* (Figure 3.11). This design will reduce bedding requirements by up to 50 per cent and will encourage feed intake because the animals are always at the correct height in relation to their feed. This design is safer for the housing of bulls than a fully bedded yard since stockmen and bulls can be kept apart. Its main drawbacks are the need for weekly or more frequent scraping and for a manure storage area, and the cost of internal fittings.

3. *Slatted* (Figure 3.12). The major advantages of slats are that bedding is saved and labour requirements are low. However, the incidence of leg and foot troubles is increased and there is evidence to show that growth rates are marginally lower than on ample straw bedding.

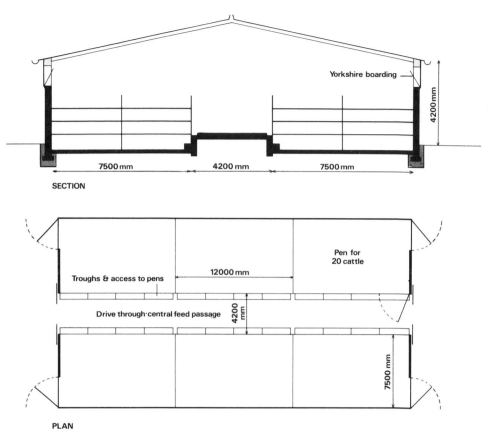

SECTION

PLAN

FIG 3.10   : FULLY BEDDED BUILDING, RAISED CENTRAL FEEDING PASSAGE – FOR OLDER STOCK

4. *Sloped floor building* (Figure 3.13). Straw usage is very low and best results are obtained when dry feeds are given. Group size must preferably be not less than fifteen small or ten large cattle. Beasts are yoked at feeding time to allow scraping of the lower part of the building.

Bull beef housing—special requirements from six to eight months onwards

The foremost concern in designing buildings for groups of young bulls must be to ensure the safety of farm staff and young children.

Ideally provision should be made for both feeding and bedding operations to be carried out from the outside of the pen. In a part-bedded house the bulls may be confined to the scraped area when bedding down and to the bedded area when scraping.

Slatted housing is particularly suited to bulls in the later stages of

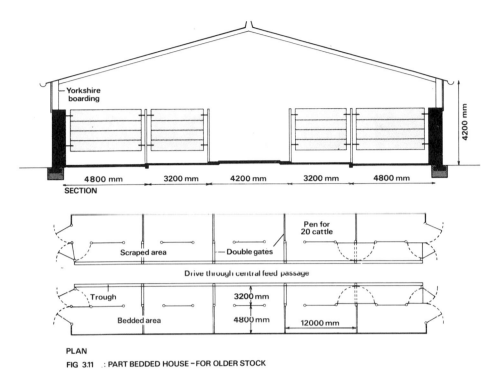

Yorkshire boarding

4200 mm

4800 mm   3200 mm   4200 mm   3200 mm   4800 mm

SECTION

Pen for
20 cattle

Scraped area   —Double gates

Drive through central feed passage

Trough

Bedded area

3200 mm

4800 mm   12000 mm

PLAN

FIG 3.11   : PART BEDDED HOUSE – FOR OLDER STOCK

finishing. Experience has shown that bulls are less active on slats than on straw, and contact between men and bulls is kept to a minimum.

Cubicles are not recommended because of the risk to bulls of injuries when mounting. The cubicles themselves may also be damaged. Fully-bedded designs are suitable provided adequate precautions are taken to ensure staff safety.

Bulls should be housed in groups of not more than twenty and provided with ample clear head room. Internal gates and pen divisions should be strong, with secure latches. Heavy duty galvanised tubular steel is most suitable and should be to a height of 1.5 m above standing level. Vertical bars have the further advantage that animals are less likely to get their heads trapped. Walls should be 215 mm thick reinforced hollow block work or 225 mm thick brickwork with piers at 2.4 m centres. Water bowls should be well protected. Space and feeding face frontages are the same as for other cattle.

All doors and gates must be well secured with childproof latches (even young bulls have a tendency to lift off unsecured gates!). Warning notices should be prominently displayed at entrances to all buildings housing bulls. Suggested wording is 'DANGER—BULLS. KEEP OUT' and must have a yellow background of at least 50 per

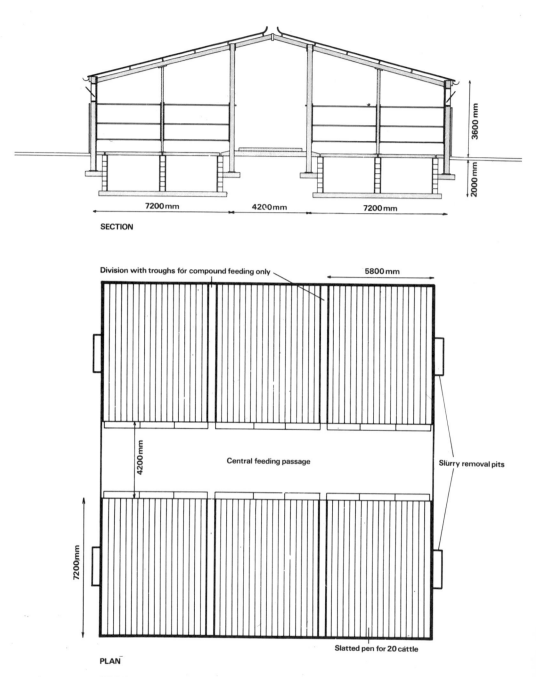

SECTION

7200 mm   4200 mm   7200 mm

3600 mm

2000 mm

Division with troughs for compound feeding only

5800 mm

4200 mm

Central feeding passage

Slurry removal pits

7200 mm

Slatted pen for 20 cattle

PLAN

FIG 3.12 : WIDE PEN SLATTED BEEF HOUSE – FOR OLDER STOCK

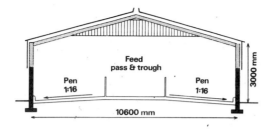

Feed
pass & trough

Pen
1:16

Pen
1:16

3000 mm

10600 mm

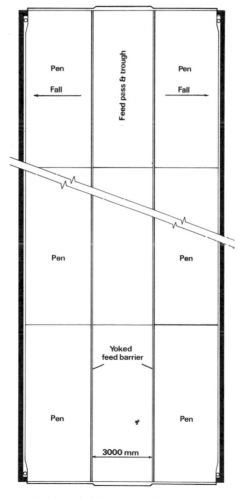

Pen

Fall

Feed pass & trough

Pen

Fall

Pen

Pen

Yoked
feed barrier

Pen

Pen

3000 mm

FIG 3.13 : SLOPED FLOOR – OLDER STOCK PREFERRED

Source : Scottish Farm Building Investigation Unit

cent of the area with a bull's head and the triangle edge in black as shown in Figure 3.14.

At least 375mm

FIG 3.14 : SAFETY SIGNS

Source : GS35 Health & Safety Executive

Any safety sign should be in accordance with British Standards BS 5378 and should comply with the Safety Signs Regulations 1980.

Advice on bull beef safety regulations should be sought from the local office of the Health and Safety Executive.

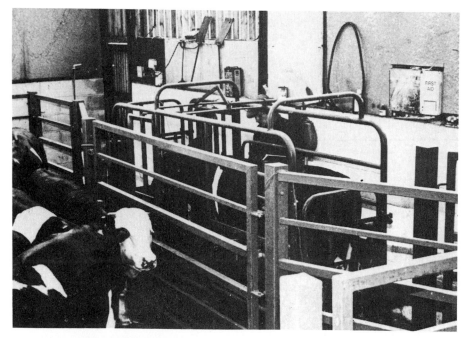

Safe and secure handling and weighing facilities for young bulls. Photograph H. F. Grundy

Cattle Handling Facilities

With any indoor system of beef production some form of cattle handling will always be necessary for routine veterinary treatment such as vaccination, foot care and delousing. Also it is strongly recommended that cattle are weighed at regular intervals so that performance can be closely monitored and feeding levels adjusted if necessary. Strong handling facilities are considered essential for bull beef production.

The essentials for cattle handling are a holding pen or pens, a forcing pen, a race, a crush/weigher and ideally, one or more holding pens for use following weighing or treatment.

The design data are as follows:

Holding pens—allow 0.95–1.4 m² per beast; walls/fences 1.53 m high.

Forcing pen—ideally should hold a minimum of fifteen beasts and have a single-sided splay with a 30° angle to form a funnel into the race. The walls or fencings of the pen should be 1.68 m high.

Race—should have a minimum length of 9 m by 0.68 m wide with 1.68 m high sides. It should hold at least five or six cattle. The race should lead to a yoke or crush or crush/weigher. Access must be allowed to all sides of the race/crush. A shedding gate should be provided at the race.

Ideally provide a roof over the crush/weigher area. The floor of the forcing pen and race should be concreted. Fences/walls should be either 48 mm diameter rails or 215 mm thick blockwork or galvanised steel tube.

The site should be free draining and a convenient loading ramp provided (Figure 3.15).

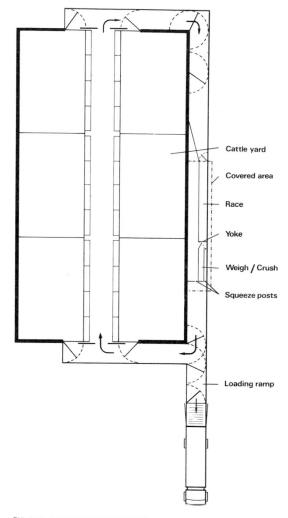

Cattle yard

Covered area

Race

Yoke

Weigh / Crush

Squeeze posts

Loading ramp

FIG 3.15 : CATTLE HANDLING LAYOUT

Chapter 4

ANIMAL HEALTH AND WELFARE INDOORS

A CODE OF RECOMMENDATIONS for the Welfare of Cattle (as provided for in Section Three of the *Agriculture (Miscellaneous Provisions) Act 1986*) has been approved by Parliament and was published in 1983. The welfare codes are intended to encourage stockkeepers to adopt the highest husbandry standards. In this way the basic health and welfare requirements of the stock may be met. These needs include:

- Comfort and shelter.
- Readily accessible fresh water and a diet to maintain the animals in full health and vigour.
- Freedom of movement.
- The company of other animals.
- The opportunity to exercise most normal patterns of behaviour.
- Natural light during the hours of daylight and artificial lighting readily available to enable the animals to be inspected at any time.
- Flooring which neither harms the animals nor causes undue strain.
- The prevention or rapid diagnosis and treatment of vice, injury, parasitic infestation and disease.
- The avoidance of unnecessary mutilation.
- Emergency arrangements to cover outbreaks of fire, the breakdown of essential mechanical services and the disruption of supplies.

The code draws attention to situations in which welfare could be placed at risk and suggests measures which can be taken to avoid this. *The Code for the Welfare of Livestock–Cattle* is published as leaflet 701 (1983) Ministry of Agriculture, Fisheries and Food; Department of Agriculture and Fisheries for Scotland; Welsh Office Agriculture Department.

Indoor systems of beef production are unlikely to offend against the codes, provided that:

1. High standards of stockmanship and feeding management are applied (see Chapters 5, 6, 7 and 8).
2. A healthy environment is maintained (see Chapter 3).
3. Diseases which may occur are promptly treated and controlled.

Stockmanship and Feeding Management

It should not be necessary to describe the elements of good stockmanship to anyone contemplating beef production, particularly in a large indoor unit. Good stockmen possess a sixth sense which enables them to anticipate problems before they arise. Close attention to the smallest details and a heartfelt concern for the welfare of animals are prerequisites.

The management and feeding required for the successful operation of the main indoor systems are specified in Chapters 5 to 8.

Environment

Details of the various building designs and of cattle handling facilities are given in Chapter 3.

The basic consideration is to ensure stock are no less comfortable than they would be outdoors where they are able to select their environment by moving from place to place.

The main considerations are temperature, relative humidity, ventilation, space requirements, structural surfaces, light and fire precautions.

Temperature

Growing and finishing beef cattle, other than very young calves, are able to tolerate quite low temperatures.

The lower critical temperatures could be defined as those at which food is used by the animal to maintain its normal temperature. The cold tolerance (lower critical temperature) of dry, well fed animals in the absence of draughts is shown in Table 4.1.

Feed intake, air speed (draughts), floor type, radiation and previous acclimatisation will all have an effect on these values. In general, on the indoor systems described, cattle are very well fed and should be fully adapted to their environment. Thus, provided they are kept free from draughts and do not lie in wet conditions, low temperatures are most unlikely to affect their performance except as very young calves.

Table 4.1 Lower critical temperatures for various classes of cattle in very low air movement conditions

	Liveweight (kg)	Lower critical temperature (°C)
Calves—new born	40	+11
—one month old	65	+7
—veal (1.5 kg DLWG)	100	−8
Young beef cattle (1.0 kg DLWG)	150	−16

DLWG = Daily Liveweight Gain
Source: A. J. F. Webster, NAC Conference 1978.

High mid-summer temperatures are a different matter and can be a problem since ruminant cattle are less able to adapt to them. Adequate shade and ventilation should therefore be provided and roofing matcrial which encourages thermal gain should be avoided.

In these circumstances roof insulation may be justified for part of the roof.

Relative Humidity (RH)
The most important undesirable effect of high RH is that the survival and spread of pathogens, particularly those of the respiratory tract, may be increased.

RH is increased by high stocking rate and by feeds of high moisture content. Poor ventilation and poor drainage exacerbate the problem. Consequently, suitable ventilation systems (see Chapter 3) and a suitable floor slope (one in twenty) with adequate bedding and drainage are required.

Ventilation
Good ventilation removes heat, moisture, noxious gasses, dust and, most importantly, air borne micro-organisms likely to cause disease of the respiratory tract.

ADAS trials at Drayton and High Mowthorpe EHFs and experiments at the Scottish Farm Buildings Unit in the late 1960s have shown that natural ventilation provides a very suitable environment for the bought-in calf for the first eight weeks. No additional benefit was obtained from controlled environment fan-ventilated housing.

Table 4.2 gives the basic design requirements for calf housing to twelve weeks of age.

British Standards 5502 quotes minimum cubic capacity at 7 m³/calf

Table 4.2   Basic ventilation design data for duo-pitch calf houses

Minimum cubic air capacity	6 m³/calf
Maximum air speed	0.2 m/sec.
Air inlet	0.045 m²/calf
Air outlet	0.04 m²/calf
Height difference (inlet/outlet)	1.5 m

NB: The values quoted are the minimum required.
Source: Dan Mitchell (1976), *Calf Housing Handbook*.

and Webster (1985) states that 10 m³/calf should be the minimum up to twelve weeks of age.

For older cattle most duo and mono pitch buildings provide satisfactory ventilation as long as suitable inlets and outlets are provided (see Chapter 3). ADAS Farm Buildings Group Technical Notes 01/1978 and 02/1979 show how the requirements can be calculated to suit individual circumstances and several specific designs are shown in Chapter 3.

Space requirements and stocking rates
The basic requirements were given in Chapter 3. In general those stated are minimum requirements. For loose-housing on straw bedding when wet rations of low dry matter silage or roots are fed, stocking rates may need to be reduced, particularly in the three to six month 'vulnerable' stage for calf pneumonia.

Table 4.3   Suggested floor space requirements when feeding low dry matter rations

Liveweight up to (kg)	Floor space requirements (m²/beast)
200	4.0
300	4.5
400	5.0
500	5.5

Table 4.3 lists floor space requirements when low dry matter rations are fed. These allowances may be considered generous, but rates in the 'follow on' (100–250 kg liveweight) houses quoted in Chapter 3, and known to give a low incidence of respiratory disease, operate at around 4.0–4.5 m²/beast. Webster (1985) now considers that provided six to eight air changes per hour occur, higher ventilation rates will not reduce pathogen contamination of the atmosphere and stocking rate becomes the key factor.

Generous air space allowance is essential to ensure temperature and humidity are close to ambient and air does not become grossly contaminated by harmful organisms. It is important that floor space should not be considered alone, and environmental air space (cubic capacity) needs to be taken into account. A. T. Smith (ADAS) recommends that the values in Table 4.4 should be used as guidelines for calculating air space.

Table 4.4   Space requirements (cubic capacity) for various age groups of stock

Age (weeks)	Liveweight (kg)	Space requirements (m³/beast)
12	100	10
20	156	12.5
28	212	15

Source: A. T. Smith, ADAS.

Structural surfaces
Floor surfaces and internal fittings must be designed to minimise the risk of injury and discomfort. All floors should be slip-resistant, well drained and durable. Slats deserve special mention in this respect. Surveys have shown that the incidence of lameness on slats, although small (under 5 per cent) is at least double that on straw bedding. Experience at Rosemaund EHF has indicated a level of 2–3 per cent lameness on slats with negligible lameness in bedded yards. Injuries on slats are normally seen in the rear feet and legs, the front legs also can be affected and large front knees are not uncommon. The construction and erection of slats should be of a high standard (grids of five to seven slats are often more accurately laid than individual ones). Because in general they provide a less comfortable and more stressful environment, slats are not recommended by the authors for cattle weighing less than 250 kg, unless straw supplies are very short.

Light
Lighting is required for adequate supervision of stock. All housed cattle should be clearly visible and twenty lux would normally be sufficient in daylight. However, fifty lux has been suggested by ADAS for satisfactory inspection at any time.

Fire precautions
Professional advice should be sought to enable stock to be evacuated quickly from buildings.

Disease in Housed Cattle Post-weaning

It could be argued that cattle continuously housed under ideal conditions and given carefully balanced rations are exposed to fewer health risks than grazing cattle. Internal parasitic infections such as parasitic gastro-enteritis and lung worm are virtually unknown. This might be expected with cattle on the cereal beef system but animals fed grass silage also appear worm-free. At Rosemaund numerous faeces samples from cattle fed continuously on grass silage (some of which had been made from previously grazed swards) have never revealed worm larvae. For this reason anthelmintic treatment is never required.

However, there are two major disease hazards associated with housed stock.

- *Nutritional disorders and laminitis*—these occur most frequently on high-energy diets and are fully dealt with in Chapters 5 to 8. Cereal beef is the system which produces the highest incidence of both ailments.

- *Bovine respiratory disease*—this is the greatest hazard of all in housed stock. It is a complex disorder involving many inter-related factors. The pathogens involved are numerous. Some of the most important are:

(a) Viruses—Respiratory Syncytial Virus (RSV) is most frequently implicated but Para-Influenza Virus (PI 3), Reovirus and Infectious Bovine Rhinotracheitis (IBR) are often involved.

(b) Mycoplasma.

(c) Bacteria—pasteurella, particularly *P. multocida*, are most frequently isolated from affected cattle and no doubt these provide the final 'death' blow.

Clinical disease in the respiratory tract would seem to be the result of a complex interaction between the pathogens, the animal, its management and its environment.

In addition, the importance of the immunity given by colostrum provided in the first few hours of life must be strongly emphasised, as must good stockmanship and husbandry.

Pneumonia is not confined to, nor greatly exacerbated by, year round housing. It most frequently occurs in the winter months (November to March) in young animals up to six months old. The most frequently affected animal is the autumn born calf of between three to six months, and the spring born calf in the first few weeks of life. Control is aided by good ventilation, sensible (i.e. low) stocking rates and good environmental conditions so that airborne pathogens are reduced. The prevention of respiratory disease may be aided by obtaining veterinary advice at the planning stage.

It is strongly recommended that stock up to six months old are housed separately from mature cattle to avoid cross infection. Also that each arrival batch is not mixed with other intakes before six months of age. The wide span building containing stock of all ages is a recipe for disaster and should be avoided. Cattle up to six months are ideally run at light stocking rates (4 m²/beast) in the type of follow-on house described in Chapter 3. However, economics rarely permit this, and a more normal rate of stocking at this stage would be 3 m²/beast. Thereafter, any of the designs for mature cattle (shown in Chapter 3) should be satisfactory as cattle are not normally affected under continuous housing conditions in later life.

Despite such measures, one cannot say categorically that control can be 100 per cent successful. The measures suggested will merely minimise the risk, and if respiratory disease breaks out veterinary advice should be immediately sought.

There are a number of additional minor ailments of housed stock mostly affecting only the younger animals.

External parasitic infections
The two most important infections are by lice and ringworm. They are an ever-present nuisance in housed stock but unless infections are severe, performance need not be adversely affected, providing treatment is carried out promptly and efficiently. Warble fly infection may very occasionally be seen in housed stock but the indoor situation should ensure it will be a very rare event.

Infestation with warble fly is now notifiable and must be reported to the Divisional Veterinary Officer.

New Forest eye
Although this corneal infection is normally seen in the grazing animal, outbreaks of this disease can occasionally be observed in housed stock, again particularly in the younger animal. Seek veterinary advice.

In conclusion, it can be said that where cattle are continuously housed in large numbers, they are undoubtedly at greater risk from respiratory disease than cattle kept outdoors. However, by extreme vigilance and the highest standards of stockmanship and veterinary attention (ideally on a regular preventive basis) these risks can be contained and stock should remain healthy throughout their lives.

Chapter 5

CEREAL BEEF PRODUCTION

Definition and Brief History

COMPOUND FEED, normally barley plus a protein, vitamin and mineral supplement is offered ad libitum from twelve weeks old to slaughter at ten to twelve months.

The system was developed from work at the Rowett Research Institute reported in the early sixties and was adopted by commercial farms in the United Kingdom.

Advantages for the producer

- Predictability—animal performance is more predictable than on forage-based systems.

- Low labour requirement—the feed is more easily handled than bulky forage.

- Carcasses—regular supplies of uniform carcasses can be produced throughout the year. Frequently these will command a premium, particularly in the summer and autumn months.

- Cash is turned over rapidly—normally in less than twelve months.

- The area of the farm does not limit the size of the enterprise.

However, there are some disadvantages

- Profitability—the system is very sensitive to changes in calf, cereal and beef prices.

- Managerial skill—a high level is required to avoid problems, especially digestive disorders and respiratory problems.

- Inflexibility—it is difficult to sell young bulls as stores part way through the system. This does not apply where steers or heifers are finished.

It is a system ideally suited to arable farms where abundant supplies of home-grown grain are available and suitable buildings and relief labour can be found without incurring additional expense. Moreover the availability throughout the year of young, tender and juicy beef with limited fat cover, is a distinct benefit to the consumer.

Animal Standard Performance Targets

Bulls are used rather than steers because they generally give a 10 per cent better growth rate and better Food Conversion Efficiency. The bull also produces a heavier carcass than the steer without becoming over fat.

The Friesian bull has proved to be ideal and targets for this animal are shown in Table 5.1.

Table 5.1   Targets for Friesian bulls

Period	Liveweight (kg)	Daily Gain (kg)	Feed Conversion (kg feed/kg gain)
3 months reared calf	100	—	—
6 months	230	1.4 (3–6 mths)	4.0:1
11 months	440	1.3 (6 mths to slaughter)	6.0:1
Overall		1.3	5.0:1

The weight of 440 kg at slaughter is based on a full on-farm liveweight. Market or overnight fasted liveweights may be some 20 kg lower, at 420 kg, giving a cold carcass weight of 235 kg and a killing out percentage of 56.

Stock Type—Bulls, Steers, Heifers

Bulls are preferred. They pose few behavioural problems on the cereal beef system due to their immaturity and relatively light weight at slaughter.

Steers, as previously mentioned (see Chapter 2), are likely to gain and convert food 10 per cent less efficiently than bulls.

Heifers of early and intermediate maturing breeds and crosses are not advisable but late maturing beef breed cross Friesian heifers (for example by Charolais or Blonde d'Aquitaine sires) are suitable. Such animals are likely to give a similar performance and carcass weight to a Hereford x Friesian steer sold at around 370 kg liveweight with an average daily gain of 0.9–1.0 kg.

Stock Type—Breeds

Early maturing breeds and their crosses (e.g. Aberdeen Angus and Hereford)
Even as bulls, these breeds and their crosses are less suited to this system as they need to be sold at very low liveweights to avoid becoming overfat.

Intermediate maturing breeds (e.g. British Friesian and Holstein)
In practice these tend to be ideal animals for the system in view of their lower calf price. Although most black and white calves from the dairy herd today are a blend of British Friesian and Canadian Holstein, distinct differences do exist between the two breed lines as shown in Table 5.2.

Table 5.2   Performance of Canadian Holstein relative
to British Friesian for barley beef

Slaughter liveweight (*kg*)	+14
Daily liveweight gain (*kg*)	−0.03
Days to slaughter	+12
Carcass weight (*kg*)	+2
Killing out percentage	−1.2
Saleable meat as percentage of carcass	−2
Saleable meat in higher priced cuts (%)	−1

Source: ADAS, Drayton & High Mowthorpe EHFs.

The Holstein type is later maturing and generally needs to be taken to higher weights than the British Friesian to obtain sufficient finish. The lower killing out percentage and poorer carcass conformation of the Holstein results in a lower yield of saleable meat. In view of this poorer potential for beef production, caution should be exercised in judging the value of Holsteins as calves for this system.

Intermediate/late (e.g. Limousin and Simmental) and late (e.g. Charolais and Blonde d'Aquitaine) maturing breeds and their crosses
These are commonly grouped under the single heading of Continental crosses, although their maturities show significant differences. Typical results are given in Table 5.3.
In general, these Continental breed crosses surpass the Friesian on this system in liveweight gain, slaughter weight and Food Conversion Efficiency. The late maturing crosses (which would also include the South Devon) are ideally suited to this system and slaughter weights of 500 kg plus are frequently reported. However, the high cost of Continental cross calves, especially Charolais and Blonde d'Aqui-

taine, needs to be balanced against premium paid for them at point of sale.

Table 5.3   Performance of a range of Continental *x* Friesian bulls

MATURITY	INTERMEDIATE/LATE		LATE	
Breed	Limousin x Friesian	Simmental x Friesian	Charolais x Friesian	Blonde d'Aquitaine x Friesian
Slaughter weight (*kg*)	450	445	470	480
Daily liveweight gain (kg)	1.3	1.3	1.4	1.4
Feed Conversion Efficiency (Friesian = 100)	115	115	113	113
Killing out percentage compared to Friesian	+2	+1	+1.5	+2

Source: North of Scotland College of Agriculture, MLC and Drayton EHF.

Effect of Varying Slaughter Weight

Current trade requirements are for a heavier cereal beef carcass. Whilst the use of bulls of later maturing breeds and crosses enables greater weights to be achieved, careful calculations should be made before animals are taken to higher weights. Not only is capital tied up longer, but more importantly feed conversion deteriorates rapidly with increasing liveweight.

This is demonstrated in Table 5.4.

Table 5.4   Effect of liveweight at slaughter on Food Conversion Efficiency

	British Friesian bulls					
Liveweight (kg)	390	410	430	450	470	490
FCE (*kg feed/kg LWG*)	6.50	7.25	8.50	10.25	10.75	11.25

Source: Frood I. J. M., University of Reading, 1976.

Management—Feeding and Health

The reared calf is gradually introduced to the 'barley beef' ration at 80–100 kg liveweight and after two to three weeks the ration is offered ad libitum.

One self-feed hopper is required per twenty beasts giving 100 mm of trough space per animal. If less space is provided competition will limit intake and this is undesirable. If more space is provided food can become stale as a result of 'mouthing over' rather than eating the ration. Clean water and roughage (usually straw) should always be

Easy feeding in a cereal beef unit. Photograph H. F. Grundy

available. The straw can be fed from racks at one kg per head per day or if plentiful clean bedding is given, sufficient will normally be obtained from the bed.

A number of health problems commonly occur in barley beef units and require speedy veterinary attention if overall performance and food conversion are not to suffer.

Bloat, acidosis and foot problems such as laminitis are associated with the ad libitum feeding of high energy diets. In such cases it may be necessary to isolate the affected animal for veterinary treatment and to feed a higher roughage diet. Re-introducing such an animal into an established group of bulls can then cause considerable behavioural problems (see Chapter 2).

Barley beef animals are particularly susceptible to acidosis or rumenitis, which may be associated with a high incidence of liver damage. The main practical steps that can be taken to minimise diet-related problems are as follows: First, ensure that a supply of palatable fresh feed is always available; secondly see that at least one kg per head of roughage is consumed daily.

Calf pneumonia can be a frequent problem in barley beef units and is likely to be the biggest single cause of mortality and morbidity. The condition is probably more severe in barley beef units than in any other system of indoor finishing. There are three main reasons for this. The first is the relatively high stocking rate in yards which is normal on this system. The second is the dry and often dusty nature of the ration which can irritate the linings of the lungs. Thirdly many farmers practise a system of continuous throughput under a single large-span roof so that different age groups of cattle share a common air space. This is likely to be conducive to pneumonia problems (see Chapter 4).

Rolling rather than grinding grain should produce a relatively dust free ration provided the moisture content is 16 per cent or above. Moist barley stored with propionic acid or in an air-tight silo is ideal. At a moisture content of 14 per cent and below grain will shatter when rolled and produce a relatively dusty ration.

On-farm methods of increasing the moisture content include soaking the barley by adding water passed through a 'constaflo' valve and into a grain auger prior to delivery to a holding bin for twenty-four hours storage. One alternative is to use a steamaliser, another is to use molasses which has the additional benefit of increasing palatability. The most important single precaution to reduce the incidence of pneumonia is to provide suitable housing (see Chapter 3).

Ration Formulation

Protein level
The following levels, based on current evidence, are recommended in the diets.

1. 90–200 kg liveweight—16 per cent crude protein (CP) in the dry matter.
2. 200–250 kg liveweight—gradually reducing to 14 per cent CP.
3. 250 kg—slaughter—gradually reducing to 12 per cent CP.

Suitable mixes are shown in Table 5.5. Fast-growing bulls may benefit, particularly in the early stages, from a 1–2 per cent higher protein level than shown above.

Protein type
A vegetable protein such as soya bean meal or animal protein such as fishmeal should be used for beasts up to 200 kg liveweight. Above this weight urea can be used in part replacement of these feeds. Below this weight, calves do not utilise urea well. At all times, it must be

thoroughly mixed into the ration to avoid urea poisoning caused by excessive intake. Urea should never form more than *one per cent* of the total diet and is best included as a proprietary grain balancer. Dried poultry manure (after checks for any pathogen presence) has been used in the mid to late stages at Gleadthorpe EHF (at a 30 per cent inclusion level) to replace all the protein. However, the use of this protein source has the effect of reducing the energy concentration of the whole ration and hence performance may suffer.

Dried lucerne and dried grass have also been used to replace all or part of the protein but, as with all protein substitutes, their prices must be carefully compared with those of other protein sources.

The protein level in barley can vary from 8 to 14 per cent depending on season and variety and hence this content should be ascertained since often in the later finishing stages rolled barley with no additional protein may be adequate.

Rations should be supplemented with the correct proportions of minerals, vitamins and trace elements.

Type of cereal and alternatives

- *Barley* (average metabolisable energy (ME) 12.9 megajoules (MJ) per kg DM; CP 100 g/kg DM). This is the most commonly used cereal. If it is more costly than other cereals on an energy and protein basis it can be wholly or partly replaced by the following:

- *Wheat* (average ME 13.5; CP 105 g/kg DM).
  Higher on average in energy and protein than barley, but significantly lower in fibre (2.2 per cent compared to 4.5 per cent). For this reason it is best fed rolled, together with approximately 10 per cent of oats and used to replace around 50 per cent of the barley. However, rolled wheat has been fed at Boxworth EHF from 130 kg liveweight to replace all the barley with no reported digestive upsets. Normal progress to slaughter was maintained.

- *Maize* (average ME 14.2; CP 85 g/kg DM).
  Higher in energy but lower in protein and fibre than barley. It can be ground through a 6 mm screen and used to replace barley partly or wholly. Roughage as hay or straw *must* always be available where maize is fed. Fat colour tends to be creamy when high levels of maize are fed.

- *Oats* (average ME 11.5; CP 105 g/kg DM).
  Similar in protein to barley but lower in energy and higher in fibre, it can replace barley wholly or partly but gain and conversion are likely to suffer and killing out percentage will be reduced. Gener-

ally, it is unlikely that its price will compensate for the lower feed value.

- *Dried sugar beet pulp* (average ME 12.5; CP 90 g/kg DM).
Slightly lower protein levels on average than barley but similar in energy, it is normally included as dried molassed pulp and is highly palatable. For this reason it is favoured for inclusion by many producers. It can provide up to 30 per cent of the ration and even up to 50 per cent of the barley can be replaced for cattle over 125 kg. It should be introduced gradually to avoid digestive upsets. With shredded or dried pulp nuts bridging and separation in the hopper may be a problem and for this reason careful attention is required.

- *Straw* (average ME 6 but very variable; CP 38 g/kg DM).
The scope for inclusion of straw in intensive diets is limited in view of its low energy and protein values and attempts to include it have rarely produced economic benefits. Without exception, it must be either ground or chemically treated if performance is not to suffer significantly, even when only a small percentage of the barley is replaced.

Scottish trials have shown that for cattle over 250 kg up to 20 per cent of the ration can be replaced by ground straw without significantly affecting performance, because the animals compensate for the lower feeding value by eating more. It has been calculated that complete diets containing 15 per cent or 30 per cent ground straw must lead to cost reductions of 33 per cent or 44 per cent before they offer any economic advantage.

Sodium Hydroxide (NaOH) treatment of chopped or milled straw can also be carried out (commercial machines are available for on-farm treatment) and using 4 to 6 per cent of NaOH will increase the digestibility of the straw. It is still not advisable to include more than 30 per cent in the ration. As well as being dangerous to handle caustic soda treatment increases water intake and urine output and hence bedding straw usage can be excessive. Urea and ammonia can also be used but as with all alkali treatments the cost is comparatively high and consequently straw cost must be very low for an economic advantage to accrue.

- *Roots*
When available, fodder beet, swedes and potatoes can wholly or partly replace barley. This is further discussed in Chapter 8.

Table 5.5 shows three suggested home mixes for intensive finishing.

Table 5.5   Examples of home mixes for intensive finishing (kg/tonne)

	Crude protein (CP) % as fed		
	16%	14%	12%
Example A			
Barley (10% CP)	775	845	900
Soya bean meal (44% CP)	200	130	75
Mineral and vitamin supplement	25	25	25
Example B			
Barley (10% CP)	600	525	600
Dried sugar beet pulp (9% CP)	200	300	300
Soya bean meal (44% CP)	150	150	75
Fishmeal (66% CP)	25	—	—
Mineral and vitamin supplement	25	25	25
Example C			
Barley	500	425	525
Dried sugar beet pulp	200	300	300
Soya bean meal	175	100	50
Maize gluten feed (20% CP)	100	150	100
Mineral and vitamin supplement	25	25	25

The inclusion of 5 per cent liquid molasses may also be recommended. To avoid problems with urinary calculi, the mineral and vitamin supplement needs to be carefully selected. Ideally an intensive beef mineral should not contain magnesium and would have at least 20 per cent salt content. For use with the mixes given above, the levels of calcium and phosphorus in Table 5.6 should be satisfactory.

Table 5.6   Levels of calcium and phosphorus for use with
mixes in Table 5.5

Mix	Calcium %	Phosphorus %
With fishmeal (Example B)	13–17	0–2
With maize gluten feed (Example C)	20–24	2–3
Others	18–22	5–7

If a relatively high phosphorus intensive beef mineral and vitamin supplement is used, ensure that the calcium content is also high.

The following levels of vitamins A, D_3 and E should be included:

> Vitamin A 150,000–200,000 iu/kg
> Vitamin D_3 15,000–20,000 iu/kg
> Vitamin E 800–1,000 iu/kg

It is beneficial to include a feed additive (See Appendix).

Current Producer Practice

Physical performance
Average and top third results recorded by MLC are given in Table 5.7.

Top third producers achieved higher gains than average and were able to take cattle to heavier weights on similar amounts of food as a result of higher than average Food Conversion Efficiency.

Table 5.7 Cereal beef results from forty groups of home reared calves

	Average	Top third
Slaughter weight (*kg*)	443	457
Days to sale	340	340
Daily gain from 12 weeks (*kg*)	1.29	1.35
Carcass weight (*kg*)	240	248
Finishing compound (*kg*)	1667	1670
Feed Conversion Efficiency (*kg feed/kg gain*)	5.01	4.81

Source: MLC 1984.

Ration formulation
Many cereal beef producers are reluctant to have more than one mix available for feeding at any one time. In these units, adjustments in formulation according to age of animal are rarely made and a constant ration of 14 per cent crude protein is fed from twelve weeks to slaughter.

Again, for simplicity protein and minerals/vitamins are provided by a pelleted 35 per cent CP grain balancer (often containing Monensin) which in small units can be easily hand mixed with the rolled barley.

Many producers feel that there are both financial and physical benefits from including up to 10 per cent of ground, pulverised or chopped barley straw. When straw is added molasses is often included to increase the palatability and consumption of the ration.

In the Eastern counties rolled wheat is now becoming a favoured ingredient because of the current lower market value of this cereal. At high wheat inclusion rates the addition of processed straw to increase the fibre content of the total diet will be beneficial.

Breeds
The Friesian bull is the preferred animal in most units. Continental cross bulls are also suitable but their high calf cost raises doubts concerning their financial viability.

Health

The greatest health hazard and a frequent problem in cereal beef units is pneumonia, particularly from three to six months of age. Mixing different age groups of stock at fairly high stocking rates under a single wide-span umbrella roof is the most important contributory factor. Farmers are constantly aware of the dangers of gaseous bloat, but many report that the incidence has been minimised since the advent of Monensin Sodium (Romensin) feed additive which reduces methane production in the rumen.

Summary

Cereal beef is considered to be the simplest of all indoor systems to operate. The feed given should be of a consistently high quality, producing the highest levels of performance in terms of animal gain and food conversion.

Labour requirements are low but financial margins are poor (see Chapter 9). There can be digestive problems associated with feeding the high-energy diet but these can be minimised by good management. Pneumonia, mainly in the young animal, is the single most important health hazard in cereal beef units.

Chapter 6

GRASS SILAGE BEEF PRODUCTION

Definition

CATTLE HOUSED throughout life and fed high quality grass silage to appetite from three months with a restricted amount of compound feed. The system is suited to the rearing of bulls, steers and heifers. Animals are finished at eleven to fifteen months old.

History

This system was developed at Rosemaund EHF, Hereford, in the late 1970s. Grazing systems of beef production were replaced by the all the year round feeding of high quality grass silage to housed cattle. This decision was based on evidence (presented in Chapters 1 and 2) which showed that:

- Modern ensilage techniques offer an attractive alternative to grazing by reducing forage wastage and thereby improving grassland utilisation efficiency.

- Performance from high quality silage can be consistently good, and checks at turn-out, at yarding and in wet weather can be avoided.

- Bulls have an inherent superiority over steers and heifers as converters of feed into beef. This potential can be exploited when bulls are kept indoors.

The feeding of high quality grass silage as the main year-round forage for bulls kept indoors has given physical and financial performance consistently superior to that of any grazing system. By allowing a degree of control over the environment and the feed regime, grass silage beef (in common with alternative indoor beef systems) has put precision into beef production.

The system has the following advantages:

- Consistently high liveweight gains.

- Efficient grassland utilisation.

- All year marketing.

- Favourable cashflow.

- Simplicity of operation.

The disadvantages are:

- High level of working capital per hectare of grass.

- Additional building requirements, particularly for silage.

- The need for additional labour. Bedding-down and cleaning out the houses are extra summer tasks. However the labour of feeding housed cattle is balanced by tasks inherent in the tending of beasts at grass. These include fencing, inspection, stock movements, anthelmintic treatment and perhaps supplementary feeding.

Performance Targets

Tables 6.1(a) and 6.1(b) show ranges of output and input targets for Hereford *x* Friesian and Friesian bulls.

Table 6.1(a)   Output targets

	Hereford x Friesian bulls	Friesian/Holstein bulls
(Fasted) saleweight (*kg*)	370–440	450–530
Days to sale from purchase (2 weeks of age)	330–385	400–470
Liveweight gain/day from 3 months (*kg*)	1.1	1.1
Carcass weight (*kg*)	200–240	240–280
Killing out percentage	54–56	53–55

Table 6.1(b)   Input targets

Compound from 3 months to sale (*kg*)	500–600	650–800
High quality grass silage (25% DM, 66 'D' value) (*t*)	3.2–4.4	4.8–6.6

Source: Rosemaund EHF, MLC, commercial producers.

Management

Healthy calves are essential to the profitability of the system. Unhealthy calves result in calf losses and inferior performance leading to unsatisfactory margins. Calf conformation and weight for age at purchase are also important indicators of an animal's potential to produce beef economically. Failure to buy calves of a good type will be reflected in slow finishing, carcasses of poor conformation, higher feed costs and a lower sale value.

It is however very difficult to spot some Holsteins as calves, but their extreme dairy type becomes more obvious as they grow. Expensive calves are not necessarily the best buy and the skilled buyer will purchase the best calves within his price range. Calf selection is of crucial importance.

The problems involved in selecting and rearing calves have persuaded many producers to buy weaned twelve week old calves weighing approximately 100 kg. This policy means that 'poor doers' have already been eliminated and a more even batch of animals may be obtained. Moreover the buying of weaned calves eliminates the high labour requirement and the risk element involved in calf rearing and enables a higher throughput of animals because a one year cycle can be achieved. However, reared calves are expensive to buy and if all the necessary facilities are available the margin per head is likely to be improved by home rearing.

The objective in calf rearing is to produce healthy well-grown calves at reasonable cost. This is achieved by the adoption of an early weaning system.

Table 6.2   Feeding bulls from weaning to sale

Time from calf arrival (weeks)	Liveweight (kg)	Feeding
5–6	65–75	Weaned: Ad lib early weaning concentrate
6–8	75–85	Calf concentrate replaced by rearing compound fed to appetite
8–12	85–105	Silage first offered to appetite
12–16	105–130	Compound gradually reduced to 2 kg/day
16–sale	130–saleweight	Silage always on offer plus 2 kg compound/day rising to 4 kg if necessary

NB: Clean water available throughout. Roughage to 12 weeks.
Source: Rosemaund EHF.

Following weaning the cattle are fed as detailed in Table 6.2.

It is essential that silage should be freely available at all times. This can be achieved by allowing stock to self-feed at the silage face but

Forage box feeding in a grass silage beef unit. Photograph H. F. Grundy

comparative trials have shown that easy-fed cattle perform better. Self-feeding often results in reduced silage dry matter intake and extra compound feed is required to maintain performance. Moreover bulls kept on a self-feed system have too much scope for restless behaviour. For these reasons silage is best easy-fed by forage box or silage block cutter.

With appropriate feeding facilities silage may be offered once daily in a quantity sufficient to last for twenty four hours. Skill is required to judge the correct quantity per pen since too little will have a detrimental effect on performance and too much will mean that silage not eaten within twenty four hours will deteriorate, resulting in loss of palatability and consequent wastage. With careful adjustment of the weight of silage offered, particularly in cool weather, it is possible to feed on six days per week only, thus easing the labour requirements at the weekend.

Since the silage face is exposed throughout the year, there is danger of aerobic deterioration, especially during the warm summer months. Silo management during feeding must therefore be directed towards minimising this loss. This can be achieved by the use of a silage block cutter to maintain a compact silage face and by the provision of relatively narrow clamp silos. When the exposed face recedes at not less than one metre per week aerobic deterioration is minimised. This is a most important factor.

The compound which replaces the calf concentrate after weaning may be a home-mix or a purchased feed formulated to supplement grass silage. Most operators prefer the simplicity and low cost of barrows and scoops for the spreading of the compound on top of the silage. The daily compound ration is normally fed half at each end of the day so that beasts can be inspected twice daily coming up to the troughs. Any lame or sick animals can then be identified. With good quality silage of at least sixty-five 'D' value, experience has shown that there is rarely any need to increase the level of compound fed beyond 2 kg per day. However any deficiencies in the quality of silage must be made good by increasing the compound ration if target gains are to be realised. This can mean feeding at least 4 kg of compound per day, particularly to late maturing cattle in the later stages of finishing.

There is now some interest in the use of complete-diet feeders. These take the form of automatic conveyor feeding systems or mobile mixer wagons. Mixer wagons are increasing in popularity and can be justified in the larger units, particularly where trough space is a limiting factor.

Breed-type

The grass silage beef system is flexible and any calf with beef potential can be successfully finished. The choice of breed-type depends on relative calf costs and on the objectives in terms of age at slaughter and carcass weights.

Although calves from the dairy herd vary in type the majority fall into one of three classes:

- Dairy breeds, mainly British Friesians and Holstein crosses.
- Early maturing beef crosses by Aberdeen Angus and Hereford bulls.
- Late maturing: beef crosses by Continental breed sires (mainly Charolais, Simmental, Limousin, Blonde d'Aquitaine) x Friesian. Also South Devon x Friesian.

Hereford x Friesians
The Hereford x Friesian bull is the preferred beast for the production of light carcasses of good conformation because of its easy management, ready finishing and swift throughput. It is an excellent converter of forage into beef with a relatively small lifetime requirement for

Hereford *x* Friesian bulls being finished on the grass silage system in part-bedded house.
Photograph H. F. Grundy

both silage and compound feed. This low silage requirement assumes particular importance in areas where the making of a large tonnage of high quality silage is difficult because of soil or climatic problems.

Hereford *x* Friesians are associated with a relatively high capital commitment/ha because of high-cost calves and their high stocking rate. They are not suitable for the production of heavy carcasses because of declining performance and over-fatness in the later stages of finishing.

Friesians and Holsteins

Demand for heavier carcasses of 250–300 kg may be met by finishing black and white beasts. On silage beef, as on other intensive systems, the Friesian has performed well. However the United Kingdom dairy herd currently contains 30 per cent of Holstein blood. Although the Holstein and the Friesian perform similarly in terms of liveweight gain and Feed Conversion Efficiency it is clear that the Holstein is an inferior beef animal in three ways. It has poorer conformation, a lower killing out percentage and a longer finishing period.

Recent experience at Rosemaund and at Crichton Royal (West of Scotland College of Agriculture) has shown that Holstein-type calves can be successfully finished on the grass silage beef system to give a

Typical fourteen-month old Friesian bull weighing around 480 kg finished on grass silage plus 2 kg compound daily at Rosemaund EHF. Photograph *Hereford Times* and *Mid-West Farmer*

slightly heavier carcass than the British Friesian. Bulls and steers of Holstein type reached a saleweight of 450 kg at Rosemaund in 370 days. Their daily gain from three months was 1.17 kg with high quality silage supplemented by 2 kg compound per day. Only in their conformation were these beasts noticeably inferior to Friesians.

Comparative results from Hereford *x* Friesian and Friesian bulls taken from MLC and ADAS surveys, MLC Beefplan and from Rosemaund show that on average, when compared with the white faced bull, the Friesian is:

• 31 kg heavier at sale;
• Fourteen days older at sale;
• 0.05 kg/day better in liveweight gain;
• A higher feed intake beast (100 kg more compound and 0.7 t more silage);
• Inferior in conformation.

Continental x Friesians

Continental beef breed crosses are noted for their well fleshed carcasses and high killing out percentage. Charolais, Simmental, and

Blonde d'Aquitaine *x* Friesians are among the latest maturing breed types and suitable for the production of carcasses weighing 300 kg or more. These continental crosses produce heavy high value carcasses. However, the calf cost is high, feed inputs large and stocking rates consequently relatively low.

At Rosemaund Hereford *x* Friesian, Friesian and Charolais *x* Friesian bulls have been compared on the silage beef system. The two runs of the trial were completed in 1983 and 1984 and involved 120 bulls in all. Half of the animals of each breed type were sold when external fat cover was estimated to be 3 or 4 L on the European Community (EEC) scale (see page 144) and the remaining half were kept until they had put on an additional 20 per cent liveweight.

When sold at the lighter weights the performance of the three breed types was similar at a gain of 1.12 kg per day from three months to sale. The Hereford *x* Friesians finished at lighter weights but had a quicker throughput and lower intakes of silage and compound.

Finishing at heavier weights revealed large differences in performance between breed types as shown in the Table 6.3(a).

Six-month Limousin *x* Friesian bulls being finished indoors on the grass silage beef system at Rosemaund EHF. Photograph *The Farmer*, Shropshire Weekly Newspapers

Table 6.3(a)   Comparison of bulls of three breed types

	Hereford x Friesian	Friesian	Charolais x Friesian
Saleweight (*kg*)	475	520	530
Days to sale from calf arrival	445	450	445
Liveweight gain/day from 3 months (*kg*)	1.0	1.1	1.15
Carcass weight (*kg*)	275	290	305
Killing out percentage	58	56	58
Feed inputs from 3 months			
Compound (*kg*)	690	710	705
Silage (25% DM) (*t*)	5.4	7.0	6.0

Source: Rosemaund EHF 1983–84.

The strengths and weaknesses of the breed types are as follows:

- The Hereford *x* Friesian bull is suitable for the production of lightweight carcasses of 200 to 230 kg. When taken to greater weights performance declines and over-fat carcasses are produced.

Six-month old Charolais *x* Friesian bulls being finished indoors on the grass silage beef system at Rosemaund EHF. Photograph I. C. Meadowcroft

- The Friesian/Holstein bull (depending on the proportion of Holstein blood) comes to sale at a carcass weight of from 230 to 300 kg.
- Charolais *x* Friesian bulls are among the most suitable breed types to give carcasses weighing at least 280 kg. They cannot be considered for early slaughter because of their relatively high calf cost.

Bulls, Steers and Heifers

Although initially developed as a system suitable for bulls, grass silage beef can be successfully operated using steers or heifers. The characteristics of these three have been compared in Chapter 2.

Bulls

The bull is clearly the top performing animal indoors in terms of liveweight gain and Feed Conversion Efficiency. It is followed by the steer with the heifer at the bottom of the performance league table. Since the bull has an advantage in performance of approximately 10 per cent over the steer and 20 per cent over the heifer it must be the first choice animal in most circumstances.

Most operators have reported an absence of management problems where Hereford *x* Friesian or continental bulls have been kept in units with appropriate facilities and it is difficult to justify the castration of bull calves of these breed-types. However, Friesian and Holstein bulls have the full dairy temperament. This means that they are less predictable in behaviour and can become unacceptably aggressive from around ten months old. This unpredictability poses hazards to stockmen during handling of the bulls and during bedding-down. If the housing and handling arrangements are not first rate or if the stockmen are not experienced and agile there is potential danger to the staff and the black and white bull should not be chosen.

Bulls and staff safety

There are statutory requirements in relation to the safety of staff tending bulls. These are explained in Chapter 2 and should be carefully studied. Although bulls generally respond very favourably to indoor conditions every effort should be made to avoid unnecessary disturbance. To this end the mixing of bulls from different pens must be avoided as far as possible (see Chapter 2) and group size must be limited.

The experiences of silage beef operators in the handling of bulls are detailed at the end of this chapter.

Stockmen must never forget that bulls are potentially dangerous and should be handled in a calm, quiet and firm manner. Where a pen of bulls must be entered two men should always be on hand.

Steers

Many silage beef units have been established on dairy farms and since bulls can be extremely restless in close proximity to cows the steer is often preferred in this situation. As detailed in Chapter 2 steers have considerable advantages over bulls in ease of management.

Heifers

The heifer is inferior to the male calf in its ability to convert feed into carcass meat. Therefore the purchase of heifer calves can only be justified by a low calf cost. They do however have advantages. Group size need not be limited and they finish readily with low feed inputs and a virtual absence of management problems.

At Rosemaund heifers have been fed silage to appetite with a supplement of one kg compound per day. Their performance is compared with that of the Hereford *x* Friesian bull in Table 6.3(b).

Table 6.3(b)   Heifer Performance

	Hereford x Friesian bulls	Hereford x Friesian heifers	Charolais x Friesian heifers
Saleweight (*kg*)	425	360	400
Days to sale from calf purchase	365	365	405
Daily liveweight gain from 3 months (*kg*)	1.1	0.8	0.9
Carcass weight (*kg*)	230	195	225
Killing out percentage	55	54	56
Feed inputs from 3 months:			
Compound (*kg*)	600	400	370
Silage (25% DM) (*t*)	4.8	4.0	4.7

Source: Rosemaund EHF.

The late maturing Charolais *x* Friesian heifer produces a much bigger carcass almost as heavy as that of the Hereford *x* Friesian bull. There is evidence to show that heifers grown at a slightly slower rate can be slaughtered at heavier weights.

Grass Conservation

The case for grass conservation is made in Chapter 1.

Hay versus silage

Most conservation is in the form of either field-cured hay or silage. From 1979 ensilage replaced hay making as the major conservation technique in the United Kingdom. There are four main reasons why more and more farmers prefer the ensilage option.

- Ensilage facilitates the intensification of grassland management by allowing increased use of nitrogenous fertilisers. It also allows easy integration of cutting and grazing.
- Ensilage is much less weather-dependent than hay-making.
- Ensilage results in earlier and more reliable aftermath growth.
- Silage is on average a higher quality and less variable feed than hay.

Table 6.4(a) shows that the silage maker can expect a conserved feed with a higher energy content and perhaps 50 per cent higher protein content than average hay. The best hays were similar in feed value to average silage, and the top-quality silages were far superior to all field-dried hays.

Table 6.4(a) The energy and protein contents of hay and silage

| | Average quality | | Top quality | |
	Hay	Silage	Hay	Silage
'D' value	56.0	64.0	64.0	74.0
ME (MJ/kg DM)	8.3	10.2	10.2	11.9
Crude protein in the DM (%)	8.7	13.6	15.0	22.7

ME = Metabolisable Energy
MJ = Megajoules
Source: ADAS Midlands and Western Region, 1985.

The hay versus silage argument may safely be declared 'no contest'. In most situations where the scale of operations warrants the necessary capital investment ensilage should be the choice.

Ensilage

The success of this system depends on high quality silage being available every day of the year.

The Grass

Good silage can be made from most types of grassland provided that the sward is well managed and fertilised. This includes long and short term leys and also permanent grass with a preponderence of productive grasses and clovers. Short term and long term leys are compared in Table 6.4(b).

Table 6.4(b) Comparison of short and long term leys for ensilage

	Perennial ryegrass/ White clover _Long term ley_	Italian ryegrass _Short term ley_
Re-seeding frequency	5–10 years	2 years
Annual yield of DM (_PRG/WC = 100_)	100	120
Sugar content of fresh forage (%)	2–4	4–6

PRG = Perennial ryegrass
WC = White clover
Source: Annual Yield—Rosemaund EHF 1980–84. Sugar Content—ADAS West Midlands
Region 1985.

Italian ryegrass has the advantages of earlier growth, higher yields (especially in its first year) and a higher sugar content. Unfortunately its productive lifespan is only eighteen to twenty-four months and therefore a frequent re-seeding cost is incurred.

For the three or four year ley the hybrid (perennial ryegrass x Italian ryegrass) grasses are ideal, and for longer leys mixtures of late flowering perennial ryegrasses and white clovers have performed well at Rosemaund. It is normally well worthwhile to include white clover is all long term leys.

Ensilage technique
This has greatly improved in recent years due to advances in mechanisation, a better understanding of the essentials and the availability of a wide range of silage additives. The process is now so well understood that the best exponents consistently make silage of high quality from year to year. There is however little margin for error. Mistakes are very costly because they result in silage of inferior feed value and a consequent requirement for larger inputs of high cost supplementary compound feeds.

Although silage analysis is not an infallible indicator of silage quality experience at Rosemaund shows it certainly gives a worthwhile guide to feed value (Table 6.5).

Table 6.5 Rosemaund average silage analysis

Dry matter (%)	25
pH	4
% in dry matter	
Crude protein	17
MAD fibre	32
Ammonia CP as % total CP	10
Estimated 'D' value	66
Estimated ME (_MJ/kg DM_)	10.5
DCP (_g/kg_)	125

MAD = Modified Acid Detergent
DCP = Digestible Crude Protein
Source: Rosemaund EHF 1983–85.

Precision chop forage harvester picking up high quality wilted grass. Photograph H. F. Grundy

This quality of silage can be consistently produced if attention is paid to the following factors:

Silage quality—Six key factors

- Cut grass early, as seed heads appear or before.
- Wilt the grass for up to twenty-four hours, if weather conditions permit.
- Use a proven additive in late season or at any time when weather conditions prevent successful wilting.
- Load the silo rapidly.
- Exclude the air by consolidation and by covering carefully with a weighted sheet.
- Minimise losses at the silage face.

The first point above *cannot be emphasised too strongly*. This is because of the need for a high energy diet for fast growing cattle. It must be remembered that the 'D' value in silage at feeding is frequently two to four units less than that in the grass at cutting. Therefore the aim must be to cut grass of at least 70 'D' value and this means at or before seed head emergence in the case of most grasses. This is shown in Table 6.6.

Table 6.6 Date of heading of a range of grasses

Hybrid RG	Augusta	8 May–22 May
Early PRG	Frances	8 May–22 May
IRG	R.v.P.	21 May–4 June
Medium PRG	Talbot	25 May–8 June
Late PRG (*tetraploid*)	Meltra	29 May–12 June
Late PRG	Melle	5 June–19 June

IRG = Italian ryegrass
Source: National Institute of Agricultural Botany (NIAB)
Farmers' leaflet No. 16—*Recommended varieties of grasses*.
NB: The date of heading is about four days earlier than 50 per cent ear emergence in a sward and will vary with season and altitude.

To obtain forage of high feed value it is advisable to cut in advance of heading. It is equally important to cut immature forage at each succeeding cut and a cutting interval of four to five weeks for Italian ryegrass and five to six weeks for long term leys has generally produced a satisfactory regrowth period. The temptation to defer cutting in order to harvest heavy grass crops must be resisted.

It is of proven importance to leave a 6–8 cm stubble particularly with Italian ryegrass and hybrids. This speeds both the wilting process and grass regrowth and reduces the chances of picking up stones and earth.

There are arguments for and against the technique of field wilting. The most thorough investigation of this in recent years was the Eurowilt programme carried out in eight countries and reported in 1984. Important conclusions were:

- Field losses of dry matter averaged 2.5 per cent for unwilted silages and 8.6 per cent for wilted silages. In-silo dry matter losses averaged 16.1 per cent for unwilted silages made with additives compared with 8.5 per cent for wilted silages made without additive. Total losses were, on average, higher with unwilted than with wilted silages.
- The silages were generally well preserved whether unwilted with additive or wilted without additive.
- Differences in feeding value were small. Wilting increased the DM

intake of growing cattle by 9 per cent but this higher intake was not always translated into improved performance.

It must be emphasised that field wilting of grass is highly desirable in that effluent production is eliminated where forage is wilted to about 28 per cent dry matter content. Wilting also increases the likelihood of a favourable lactic fermentation. It should however be noted that prolonged wilting increases field losses and the aim should be a wilting period of less than twenty-four hours. Weather permitting the resulting dry matter content will be 25–28 per cent. When showery conditions prevent effective wilting, silage quality is ensured by using a proven additive. This is preferable to the halting of ensilage until the return of favourable weather or the extension of the wilting period. Meter chop forage harvesters tend to promote a good fermentation by an optimal release of sugar from the herbage because of their short chopping action. This short material is also readily handled out of the clamp and may increase intake. However meter chop machines are not essential and good silage is made using double chop harvesters. Flail harvesters are used for direct cutting—but beware of effluent.

The fast loading of the silo and careful sheet management are important in limiting in-silo losses. The presence of top and side waste in a clamp silo is indefensible. Top waste is avoided by ensuring good contact between the weighted top sheet and the top surface of the grass. Shoulder waste is the most difficult to eliminate but can be minimised by folding the side sheets on to the top and overlapping with the weighted top sheet.

Silage effluent is potentially a potent pollutant of water courses and its escape must be prevented by the installation of an effective collection system. It is recommended that all silage makers take note of the provisions of the *Control of Pollution Act 1974*.

Planning the silage supply
The provision of high quality silage in adequate quantity every day must be carefully planned.

The requirement of 25 per cent dry matter silage is approximately 4–5 t/head for Hereford *x* Friesian bulls or steers and 5.5–6.5 t/head for Friesians and Continental *x* Friesians. The silo space required must be carefully calculated before embarking on the system (see Table 6.7).

When satisfied that the total supply is adequate the operator must then ensure its continuous availability. This normally implies a carry-over from the previous season's supply and hence the need for more than one silo.

Table 6.7 Silo space requirements

Breed type	Silo space/beast (cubic metres)
Hereford x Friesian	5–7
Friesians and Continental x beasts up to 300 kg carcass weight	8.5–10
Beasts taken to over 300 kg carcass weight	10+

Based on 660 kg of 25 per cent DM silage occupying one m³.

In a unit feeding 1,000 t of silage from two cuts taken per annum two silos could be managed as in Figure 6.1.

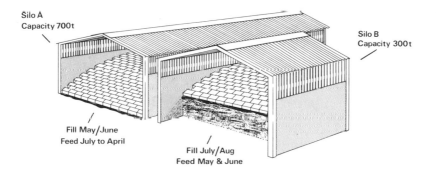

Silo A
Capacity 700t

Silo B
Capacity 300t

Fill May/June
Feed July to April

Fill July/Aug
Feed May & June

FIG 6.1   :THE USE OF TWO SILOS IN A SILAGE BEEF UNIT

Source: Rosemaund EHF

This is a very simple example, and many operators take three or more cuts. Where this is the case the availability of three or more silos gives more flexibility, and allows the ideal arrangement of one silo for each cut. If there are no more than two silos and several cuts are taken, the unavoidable 'joins' between forage from different cuts may lead to higher in-silo losses.

Silage made in big bales may perform the function of silo B in filling the gap between one season's supply of clamp silage and the next, and one clamp silo plus a supply of big bales may be the minimum requirement to ensure continuity of supply. Big bale ensilage may thus be an auxiliary system of silage making. However resultant material may not in all cases be as reliable as clamp silage in its consistency of quality or in its shelf-life.

The Compound Ration

Silage fed on its own (however excellent) is unlikely to produce liveweight gains at the target level of one kg/day from males or the 0.8 kg/day required from heifers.

Young beef cattle on all-silage diets have usually gained between 0.7 and 0.9 kg/day. The supplementary feed must contain energy in a concentrated form, necessary minerals and vitamins, and also high quality protein.

Protein supplementation

This complex subject has been well summarised by ADAS nutrition chemists as follows. Good quality grass silage has a high CP and digestible crude protein (DCP) content. However research has shown that the conventional measurement of DCP content of a feed does not necessarily give a true measure of its protein value for ruminants. DCP provides no indication of the amount of crude protein that is degraded in the rumen (rumen degradable protein or RDP), nor is it an indicator of non-degraded protein (UDP) or of its composition.

Grass silage is low in UDP and where small amounts of soya bean meal or white fishmeal have been included in the compound fed to supplement a grass silage diet, growth responses have been observed. This has been explained as a response to supplementary UDP.

The response to two levels of white fishmeal in the compound offered to silage-fed Friesian bulls at Rosemaund is shown in Table 6.8.

Table 6.8   Response to white fishmeal supplementation of silage/barley ration

Compound/day	Weight at 4½ months (kg)	Weight at 9 months (kg)	Daily liveweight gain (kg)
2 kg mineralised barley	168	334	1.06
1.9 kg min. barley + 100 g WFM	168	340	1.12
1.8 kg min. barley + 200 g WFM	167	354	1.20

WFM = White fishmeal
Source: Rosemaund EHF 1982–83.

There are a number of alternative protein feeds high in UDP. However, Rosemaund experience favours the inclusion of 200 g of white fishmeal in the compound to nine months of age or 350 kg liveweight. Some trials elsewhere have not supported this finding.

In the later stages of finishing the effect of this inclusion is less certain. Results are conflicting, but generally do not support the use of fishmeal for heavy animals.

Mineral/Vitamin supplements

Supplements containing all the necessary minerals and vitamins for cattle are readily available and should be used at the recommended inclusion rates as detailed in Table 6.9. When fishmeal is used these rates can be considerably reduced.

Table 6.9 Recommended mineral/ vitamin supplement

Calcium %	20
Phosphorus %	6
Magnesium %	2
Salt %	20
Manganese (mg/kg)	4,000
Zinc (mg/kg)	2,000
Copper (mg/kg)	800
Cobalt (mg/kg)	100
Iodine (mg/kg)	200
Selenium (mg/kg)	5
Vit A (iu/kg)	250,000
Vit D_3 (iu/kg)	65,000
Vit E (iu/kg)	250

The recommended inclusion rate for this supplement is 25 kg/tonne of the compound ration.
Source: ADAS.

Clean water should be freely available at all times.

There may well be scope for a reduction in the cost of the compound fed by using a wide range of substitute feeds including the various cereal grains and arable by-products. In the future, rations may be periodically re-formulated on the 'least cost' principle but care should be taken to ensure that feed substitutions do not cause digestive disturbances or nutritional deficiencies and thereby reduced performance.

Animal Health

The maintenance of good health in housed cattle is discussed in Chapter 4.

On this system stock health has generally been very good. As in all beef systems which involve the housing of cattle, lung disease may occur. This is a complex problem discussed in Chapters 3 and 4.

On-farm Performance

Average performances achieved at Rosemaund and at the National Agricultural Centre (NAC) unit are shown in the Table 6.10. Results from three surveys of commercial units carried out by ADAS and MLC are also included. All figures are for Friesian bulls. Comparable results from other breed-types and from heifers are presented in Tables 6.3(a) and 6.3(b), pages 73 and 75.

Table 6.10　　Silage beef performance—Friesian bulls

	Rosemaund EHF (1980–84)	NAC (1982–83)	MLC Beefplan (1983–84)	MLC/ADAS survey (reported 1983)	ADAS Northern Region (1983–84)
Saleweight (kg)	449 (fasted)	526 (full)	498 (full)	487 (full)	473 (full)
Days to sale from 3 months	290	390	377	360	363
DLWG from 3 months (kg)	1.09	1.00	1.02	1.02	1.00
Feed inputs from 3 months compound (kg)	688	1206	1016	1080	978
Silage (25% DM) (t)	5.0	6.5	5.8	5.6	5.7

DLWG = daily liveweight gain

Typical thirteen-month old Hereford *x* Friesian bull weighing around 440 kg finished on grass silage plus 2 kg compound daily at Rosemaund EHF. Photograph *Hereford Times* and *Mid-West Farmer*

It is interesting to compare performance with the targets originally set. The majority of units have achieved a daily liveweight gain in excess of one kg from three months and have sold within the target weight range of 450–530 kg. However most operators have chosen to ensure rapid progress and a swift throughput by feeding from 3 to 4 kg of compound each day in the later stages. This has resulted in a compound input of around one tonne per beast in many units. This level of feeding could be reduced (with considerable cost saving) by the production of higher quality silage.

The Experience of Commercial Operators

A questionnaire completed by twenty-five leading silage-beef producers yielded the following points of interest.

- Many aim at the production of carcasses of over 250 kg and there-fore operate with Continental crosses or Friesians. A few prefer the rapid throughput achievable with early maturing cattle, a notable example being King Brothers, Lane Ends Farm, Salterforth, Lanca shire. King Brothers find that Hereford x Friesian bulls can be finished at relatively light weights with a silage intake of well under

Normal daily ration for thirteen-month old Hereford x Friesian bull fed high quality grass silage.
Photograph *Hereford Times* and *Mid-West Farmer*

4 t/beast with a consequently high stocking rate of 12½ beasts/ hectare. This minimises the production of unwanted fat as bulls are finished at around 370 kg (carcass weight 210 kg) in about ten months.

- Most rear and finish entire cattle because of their superior perform-ance. Friesian and Holstein bulls are the only breed types which have given cause for concern because of aggressive behaviour from ten months of age onwards. The following factors have pre-disposed to calmness in the bull beef unit.

  * the use of strongly-constructed, well-designed buildings and handling facilities.

  * the establishment of settled groups of around twenty bulls from weaning. However operators have reported satisfaction with the behaviour of bulls in group sizes of up to thirty-three.

  * a strict 'no-mixing' policy between pens.

  * the avoidance of all unnecessary contact between staff and bulls. Therefore bulls are handled and weighed only when really necessary and are bedded down no more frequently than needed to keep the animals clean and comfortable.

- The main health problems reported were lung disease, ring-worm, lice infestation and New Forest disease (inflammation of the conjunctiva of the eye). Bloat and scour were less important health hazards. Lameness (affecting around 2 per cent of the beasts) was reported amongst cattle kept on slatted floors. This occurred mainly when beasts were moved from a bedded house to a slatted floor house.

- Many producers aim to make a 65 to 70 'D' value silage. This implies in some seasons the continuation of cutting and ensilage during showery weather, a policy which cannot always be followed on some heavier soils because of the likelihood of machinery damaging the swards. Since the average 'D' value of British silages is sixty-two it is clear that silage beef operators make a well above average quality feed. The majority of producers reported that aerobic deterioration at the silage face was well under control.

- Many operators feed 3 or 4 kg of supplementary compound per beast daily in the later stages of finishing in order to ensure that an adequate level of finish is achieved.

- The labour requirements are not considered to be onerous. However the high level of working capital is a worry to producers with high bank borrowings.

Summary

It is clear that the system fits in best where:

- Suitable beef houses and silos are available on the farm.
- Healthy calves can be reared or purchased with low calf losses.
- High quality (65 + 'D' value) silage can be consistently made.

A well managed silage beef unit may be financially a reasonable alternative to dairying, although capital requirements are high and management failures exact a severe penalty.

Chapter 7

MAIZE SILAGE BEEF PRODUCTION

Definition

HOUSED CATTLE fed maize silage to appetite together with a suitable ration of supplementary high protein compound. Bulls, steers or heifers are finished at twelve to sixteen months of age.

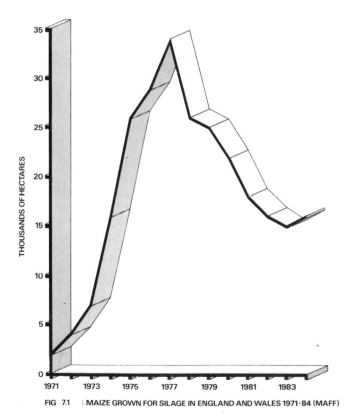

FIG 7.1 : MAIZE GROWN FOR SILAGE IN ENGLAND AND WALES 1971-84 (MAFF)

History

Maize has been grown in Europe since the sixteenth century and in the United Kingdom since the early nineteenth century. The breeding of hybrid varieties of the crop has brought advantages in terms of enhanced cold tolerance, higher yields and earlier maturity. In the period 1970 to 1977 there was a big increase in the area grown for feeding to beef cattle. However poor establishment, particularly when grown in less suitable areas, and the problems brought about by its late harvest led to a contraction of the area grown. With the recent introduction of earlier maturing cultivars the popularity of maize is once again on the up-turn (see Figure 7.1).

Why Forage Maize?

Advantages

- When grown in a favourable situation, modern varieties can reliably produce a big bulk of forage of high energy content.

- Maize can be grown continuously (up to fourteen consecutive years recorded in the United Kingdom) until weed problems supervene.

- Maize is an excellent break crop, allowing the control of grass weeds and of most pests and diseases of cereal crops. It also improves soil structure (but not in wet autumns). The late spring sowing date gives time for thorough cultivations. Alternatively it can follow a crop of forage rye (as demonstrated at the GRI) or follow a spring grazed grass ley.

- Prior to the sowing of maize land can be used as a slurry dumping ground. This practice can lead to savings in the fertiliser costs of growing the crop.

- Maize requires cutting once only. For this reason it can be regarded as a contractors' crop and capital expenditure on sowing and harvesting machinery may be avoided.

- Unlike grass, the 'D' value of maize changes little during the later stages of development. Cutting date is therefore not as critical. The ensilage of maize poses no particular problems and there is no requirement for an additive.

- With its consistently high energy content and good palatability forage maize silage is an excellent feed for beef cattle.

Disadvantages

- The growing of forage maize is restricted to areas where good yields are reliably obtained. In marginal areas yields can be disastrous in some years. Moreover, forward feed planning is difficult until the crop is in the silo and the yield known.

- On heavy land the late harvest brings difficulties in some seasons and can clash with other arable field operations, in particular with the harvesting of sugar beet and potatoes. The harvesting of forage maize may delay the seeding of a following crop of winter wheat until after the optimal sowing date.

- Forage maize silage can be subject to aerobic deterioration at the silage face.

- Maize silage is low in protein content and low in most minerals.

Where can Forage Maize be Grown?

For optimum yields adequate moisture and warm sunny conditions are essential during its five months growing period. Warmth is frequently the limiting factor and the Table 7.1 shows the requirements in terms of Ontario Units. Since maize shows little or no growth below 10°C, the Ontario Heat Unit system was developed where the lower limit is 10°C. Cumulative temperature is calculated from a mean maximum and minimum temperature minus 10°C multiplied by time in days.

Table 7.1   Ontario units and accumulated temperature for different crop stages in forage maize

Crop stages	Ontario units	Approx. equivalent (Day degrees above 10°C)
From sowing to 25% whole crop dry matter	2300	660
From sowing to 30% whole crop dry matter	2500	750

Source: MAFF Booklet 2380.

Consistently good crops of maize are most likely to be grown south of a line from Bristol to the Wash. In favourable years high yields may be obtained in the Midlands and as far north as York. It should be remembered that even in the favoured south eastern area of England many sites are unsuitable, particularly those at an altitude

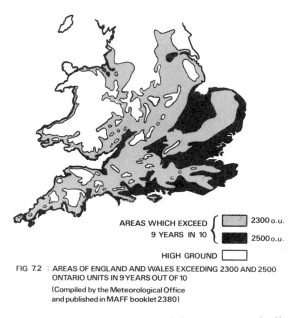

AREAS WHICH EXCEED { ▨ 2300 o.u.
9 YEARS IN 10 { ▉ 2500 o.u.

HIGH GROUND ☐

FIG 7.2 : AREAS OF ENGLAND AND WALES EXCEEDING 2300 AND 2500
ONTARIO UNITS IN 9 YEARS OUT OF 10
(Compiled by the Meteorological Office
and published in MAFF booklet 2380)

of over 150 m, in a cold exposed position, or on shallow soils (see Figure 7.2).

Comparison with grass silage

Yield
Forage maize has been characterised by variability of yield but this is less true with modern varieties.

Most crops yield between 7 and 15 t/ha/annum DM with an average of around 10 t/ha/annum. Some National Institute of Agricultural Botany (NIAB) trials have averaged over 12 t/ha/annum.

Modern varieties of forage maize are generally harvested in late September or early October when the crop has reached the 25–30 per cent DM stage. The tendency is to aim for the top of this dry matter range in order to eliminate the production of silage effluent.

Maize should be cut when the grain is at the firm cheesy stage. The cutting date is not critical because digestibility falls little over a three or four week period at harvest time. This is a big advantage since it allows harvesting to be delayed if necessary until a spell of good weather or until the dry matter content of the crop increases to 25–30 per cent.

Specialised silage maize harvesters are available, but precision chop forage harvesters fitted with maize cutting attachments are most widely used. Chopping of the crop to a length of 60–120 mm maximum is advisable to promote a rapid lactic fermentation and to facilitate consolidation of the forage in the silo. A suitable fermentation and low in-silo losses are favoured by fast filling using the wedge technique, consolidation of the clamp (more important than with grass silage) and careful polythene sheet management to secure an air-tight silo. The top sheet should be stretched tight over the surface of the maize and well weighted.

There is no requirement for the use of a silage additive since maize forage contains plenty of sugar which ensures a rapid rise in lactic acid content in the silage.

Although the ensilage of forage maize poses few problems *there can be considerable losses during the feeding period*. This problem is soluble however and is discussed in the management section on page 96.

In calculating silage stocks and silo space requirements it should be remembered that 28 per cent DM maize silage settled to a depth of 2 m has an approximate density of 700 kg/m³. It is slightly more dense than an average grass silage (please see Table 6.7 on page 81).

First year Italian ryegrass leys are capable of yielding 15 to 20 t/ha/annum of dry matter. Long-term perennial ryegrass/white clover leys at nineteen sites in the United Kingdom averaged almost 11 t/ha/annum of DM when cut six times annually from 1977 to 1981 (National Grassland Manuring Trial 23). Most grass silage beef producers grow both short and long term leys and perhaps some permanent grass, and achieve silage DM yields of 8 to 10 t/ha/annum plus some late season grazing.

Quality

A comparison of the composition and nutritive value of maize silage as opposed to grass silage is made in Table 7.2.

Table 7.2 **Typical composition and nutritive value of maize silage and grass silage**

	Maize silage	Grass silage (medium quality)
Dry matter (%)	27.0	25.0
Crude protein (% in DM)	9.0	14.0
pH	3.8	4.2
'D' value	68.0	63.0
Estimated ME (MJ/kg DM)	11.0	10.0
Estimated DCP (g/kg DM)	50.0	90.0

Source: MAFF Booklet 2380.

There is a wide range around these figures. For instance modern maize varieties frequently make 30 per cent DM silage. The majority of samples of maize silage give an analysis of 10.7 to 11.1 ME (68 to 70 'D' value). Only a minority of grass silages have an energy content as high as this. Forage maize is a high energy feed equalled in this respect by only the very best grass silages made from leys cut at the pre-flowering stage.

Grass silages, however, have a higher protein content than maize silage with the DCP ranging from 80 to 150 g/kg DM compared to the 50 to 60 g/kg DM normal in maize silage. Maize silage also contains inadequate amounts of several minerals and of vitamin E (details later in this chapter on page 98).

Direct growing costs

Table 7.3(a)   Forage maize and grass:
variable costs (£) per hectare and per tonne DM (1986)

Input	Forage Maize	Grass (two year ryegrass ley)
Seed (£)	74	20 (annual seed cost)
Fertiliser* (£)	76	205
Weed control (£)	10	11
Pest control (£)	25	—
Total variable costs (£)	185	236
Yield of DM (t/ha)	10	10
Total variable costs (£/tonne DM)	18.5	23.6

* Fertilizer composed of 60N, 75P$_2$O$_5$, 150K$_2$O kg/ha for forage maize and 375N, 100P$_2$O$_5$, 150K$_2$O kg/ha for grass.

The fertiliser costs of growing forage maize are much less than those incurred in growing grass. In experiments in southern England maize has been relatively unresponsive to nitrogen, phosphorus and potassium (NPK) and where the crop is grown after cereals 60 kg/ha N has been found satisfactory. Following slurry applications levels of NPK fertilisers may be reduced below those in Table 7.3(a) but local advice should be sought. Despite the cost of pest control maize forage may be £5/t DM cheaper than grass forage. In both cases the costs of ensilage machinery (whether owned or belonging to a contractor) and of silos must be added to calculate the cost of silage production.

There are large scale maize growers who limit their ensilage machinery costs to one half of the figure shown in Table 7.3(b). Likewise the machinery costs of grass ensilage may be much reduced by large scale operations, co-operative use of machinery or taking

Table 7.3(b)   Silage costs per tonne DM 1985 (£)—(excluding land and labour costs)

Costs	Maize silage	Grass silage
Growing of forage	18.5	23.6
Ensilage machinery (including maintenance)	16.8	32.4
Silo (including sheeting)	16.0	16.0
Silage additive	—	4.0
Cost per tonne DM	51.3	76.0
Cost per tonne silage (25% DM)	12.8	19.0

only one or two cuts. Machinery costs per cut are normally higher for maize than for grass but over the season these costs may be double or treble for the grass ensiler compared to the forage maize grower who cuts only once.

Also, it has been assumed in Table 7.3(b) that half the grass silage is made with the aid of an additive.

Therefore the evidence suggests that:

- Forage maize can equal or surpass grass leys as a producer of high energy forage.

- Maize silage can be produced at a lower cost than grass silage.

However it is important to remember that these advantages will *only exist in areas where maize yields consistently well.*

Performance Targets

The beef unit at the National Agricultural Centre, Stoneleigh has been prominent in the development of a maize silage beef system. Tables 7.4(a) and (b) show the output and input targets set by the beef unit.

Table 7.4(a)   Output targets—Friesian bulls

	kg
Purchase weight	45
12 week weight	105
Saleweight (full)	495
Carcass weight	255
Days to sale from calf purchase	450
Liveweight gain/day 12 weeks to sale	1.1
Liveweight gain/day overall	1.0

Table 7.4(b)   Input targets to sale—Friesian bulls

Maize silage dry matter from 12 weeks (*t*)	1.8
Protein concentrate from 12 weeks (*kg*)	500
Rolled barley from 13 months (*kg*)	150
Stocking rate (based on silage yield of 10 t DM/ha)	5.7 beasts/ha

Source: Beef unit NAC Stoneleigh.

Feeding Management

Table 7.5   Feeding from three months to sale

Time from calf arrival	Liveweight (Full) (kg)	Feeding
3–13 months	105–430	Maize silage to appetite + 1.5 kg/day of 34% CP protein/mineral/vitamin supplement
13–15 months	430–490	As above + 2 kg/day of rolled barley

Source: Beef Unit NAC Stoneleigh.

The target liveweight gain from three months to sale is 1.1. kg/day on the system shown in Table 7.5. The evidence available from feeding trials at the Grassland Research Institute, the National Agricultural Centre, and from Boxworth and High Mowthorpe EHFs is that maize silage alone cannot supply sufficient protein to achieve this target. The flat rate feeding of 1.5 kg/day of a protein/mineral/vitamin supplement from three months to sale is now practised in many units. This is well illustrated in Figure 7.3 taken from *Farmers Weekly* 19 October 1984.

Since silage intake increases throughout the life of the beast the proportion of protein in the dry matter of the complete diet diminishes

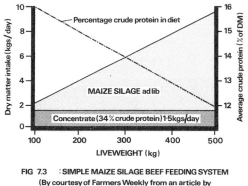

FIG 7.3   : SIMPLE MAIZE SILAGE BEEF FEEDING SYSTEM
(By courtesy of Farmers Weekly from an article by
Dr. Mike Wilkinson, 19 October 1984)

gradually in accordance with the needs of the animal as can be seen in Table 7.6.

Table 7.6   **Protein content of the diet from 100 kg**
liveweight to sale

Liveweight (kg)	CP in DM of whole diet (%)
100 to 200	16
200 to 300	14
300 to sale	12

Feeding the silage

Maize silage should be gradually introduced and then be continuously available. Self feeding has proved to be unsuitable because of increased wastage at the silage face. The material is very easy to handle out of the clamp and easy feeding is preferred. However it must be emphasised that maize silage needs careful management at the clamp face as it is an ideal medium for mould growth, particularly in warm weather. Moulds have been detected up to two metres into the clamp from the face and it is important to cut back the face at least one metre a week. A narrow clamp is a great help in achieving this. However, good compaction in the silo is all important in limiting aerobic deterioration and is promoted by careful consolidation during filling together with the maintenance of a tight feeding face during emptying.

Maize silage should normally be fed daily, since less frequent feeding will mean feed wastage through deterioration and lowered animal intake, as palatability is reduced by heating and mould growth.

The compound ration

Trials at the GRI have shown that in the three to four months age range animal or vegetable protein must be fed rather than urea. Otherwise liveweight gain will be lowered. From six months (200 kg liveweight) urea can replace protein in the concentrate with a consequent saving in the cost of the ration. However urea must always be fed with care, the inclusion rate in the diet not to exceed one per cent of the total feed.

Where all the animals can feed at the same time the compound ration may be offered in two feeds daily on top of the silage. However many maize beef producers operate complete feeders. These allow changes in the make-up of the ration with little danger of feed rejections. Mixer wagons also allow a reduction in the manger frontage per beast from the normal 550 to 700 mm for a 350 kg beast to as little as 175 mm per beast.

Feeding by mixer waggon in a maize silage beef unit. Photograph *Hereford Times* and *Mid-West Farmer*

Using the 'flat rate' feeding system Friesian bulls should gain over one kg per day from around five months old and should weigh approximately 410 to 430 kg at twelve months. Producers aiming at a relatively light saleweight of around 450 kg should achieve this at around thirteen months with no change in the feed regime. Alternatively bulls may be finished at 480 to 500 kg at fourteen to fifteen months of age with total feed inputs per head from three months to sale of 6 to 7 t of maize silage plus 500 kg of protein concentrate plus 150 kg of rolled barley fed in the later stages.

With lower DM silage extra compound will be required to maintain the target level of liveweight gain. For instance an additional 2 kg of compound per day may be needed with maize silage of 20 per cent DM.

It is important to note that, compared to grass silage, maize silage is low in phosphorus, sodium, copper, manganese, zinc and vitamin E. Therefore when ordering a mineral/vitamin supplement (whether on its own or incorporated in a concentrate) it must be made clear that it is to be fed with maize silage. The analysis of the supplement should approximate to that shown in Table 7.7.

**Table 7.7   Recommended mineral/
vitamin supplement**

Calcium (%)	20
Phosphorus (%)	10
Magnesium (%)	2
Salt (%)	30
Manganese (*mg/kg*)	4,000
Zinc (*mg/kg*)	2,000
Copper (*mg/kg*)	1,000
Cobalt (*mg/kg*)	100
Iodine (*mg/kg*)	200
Selenium (*mg/kg*)	5
Vit A (*iu/kg*)	250,000
Vit D3 (*iu/kg*)	65,000
Vit E (*iu/kg*)	5,000

The recommended inclusion rate is 25 kg of the supplement per tonne of the compound ration.
Source: ADAS.

Type of Animal

The superior performance of the bull (see Chapter 2) may be fully realised on this system, and the bull is the preferred animal in most units. In situations where bulls cannot be tolerated steers or heifers may be finished, in which case the return per beast will be lower because lighter carcasses will be sold. However, compound inputs are likely to be less. In common with other intensive beef systems, later maturing breed types are preferred and the Friesian is an appropriate and frequent choice. Continental *x* Friesian calves are widely used to produce carcasses in excess of 250 kg. Hereford and Aberdeen Angus cross beasts (particularly bulls) may be used for earlier selling at relatively light carcass weights up to 230 kg.

Current Producer Practice

There is a trend towards monoculture of the forage maize crop since this ensures that it is always grown on the most suitable fields.

Results from maize silage beef units are typified by those reported and shown in Table 7.8 from the Grassland Research Institute, Boxworth EHF, and the National Agricultural Centre beef unit.

Many producers are now finishing bulls at 500 kg, at fourteen to fifteen months of age. During the final two to three months 2–3 kg of rolled barley per day are fed in addition to the protein concentrate to maintain performance and ensure the production of carcasses with sufficient fat cover and giving good conformation scores.

Table 7.8   Maize silage beef performance

	GRI (1973/74) Friesian bulls	Boxworth EHF (1975/76) Friesian steers	NAC Beef Unit (1978/79) Friesian bulls
Initial weight (*kg*)	100	111	92
Final weight (*kg*)	452	446	496
Age at slaughter (*months*)	15	14	15
DLWG 12 weeks to sale (*kg*)	1.0	1.1	1.1
Total concentrate input (*DM t/head*)	0.53	0.53	0.62

Sources: Wilkinson J. M. & Penning I. M. 1976 Animal Prod: 23, 181–190. Bowerman P. 1977 Animal Prod: 24, 154. Read D. J. 1980 Proc. 1st European Maize Congress, Cambridge.

A good example of the way in which the use of a complete feeder facilitates changes in feed formulation is provided by the beef unit of Mr Richard Vaughan, Huntsham Court, Goodrich, Ross-on-Wye in Herefordshire. In this unit the proportions of maize silage and compound in the feed are varied according to the stage of growth of the bulls. All ration changes are made gradually in line with the programme in Table 7.9.

Table 7.9   Changes in ration formulation

Liveweight (kg)	Feed (dry matter)
100 to 200	30% silage, 70% compound
200 to 350	80% silage, 20% compound
350 to sale	23% silage, 77% compound

Source: Mr R. Vaughan.

In the three to six months age range the bulls are given a high proportion of compound feed to encourage a good start at a time of life when the feed conversion ratio is favourable. From six to nine or ten months silage is the main feed and with increasing intake in this period good liveweight gains are obtained at low cost. In the later stages the proportion of compound feed is again increased to ensure adequate finish and good carcass conformation. The higher level of barley feeding is reported to produce carcasses with a pale fat colour only a little creamier than that of cereal beef carcasses, and this has been reflected in the high sale prices obtained.

The approach adopted in this Herefordshire unit has given excellent results. Friesian bulls are finished at 450 kg in twelve months with feed inputs per head of 3.8 tonnes of maize silage and 1.3 tonnes of compound. The supplementary compound costs are high, but there is a swift throughput and an excellent stocking rate of over eleven beasts per hectare.

Some producers feed both maize and grass silages. This may be done in two ways. If stored in separate silos it is normal for the maize silage to be fed in the winter and grass silage (which is less subject to heating and deterioration in warm weather) in the summer. Alternatively maize may be loaded into the silo on top of grass silage to give a mixed feed. The higher protein content of the grass silage brings economies in the feeding of supplementary protein concentrates and the mixture of forages is said to improve the palatability, resulting in increased intakes.

Summary

Maize silage is a palatable high energy feed. Where soil and climatic conditions are favourable high yields of maize can consistently be grown at a lower cost per tonne of DM conserved than grass silage.

Plant breeders are continuing to improve varieties resulting in better growth during the critical spring period and in earlier maturity. Success here would improve the competitive position of forage maize.

The necessary cautionary note is that maize yields tend to be more variable from season to season than those of grassland. However this tendency is now less pronounced where maize is grown in acceptable areas with the benefit of up-to-date varieties and techniques. Nevertheless considerations of soil, topography and climate impose limitations on the adoption of this exciting crop to produce forage for the beef unit.

Chapter 8

OTHER ALTERNATIVES FOR INDOOR FEEDING

Modifications of the Rosemaund Silage Beef System

ON THIS SYSTEM the typical feed regime is high quality grass silage to appetite with supplementary compound fed at 2 kg per head per day. However the effects of increased supplementary feeding have been studied at the Crichton Royal Farm of the West of Scotland Agricultural College and at Rosemaund EHF. At Crichton Royal Dr Jerry Hughes introduced ten week old Holstein/Friesian bull calves to silage, and fed in addition, 3–3.5 kg/day of compound to achieve a finish at twelve months of age. In three Rosemaund trials Hereford x Friesian bulls were fed silage supplemented with 2 kg compound/ day from ten weeks to between thirty-nine and forty-four weeks. Thereafter until sale they were allowed 4 kg/day and ad lib silage (Table 8.1).

Table 8.1   Performance of bulls fed silage and up to 4 kg compound/day

	Rosemaund EHF 1979–81	Crichton Royal Farm 1982–83
Number of bulls	32	24
Daily liveweight gain (start to sale) (*kg*)	1.03	1.06
Saleweight (RM fasted, CR full) (*kg*)	428	454
Days to sale	368	383
Carcass weight (*kg*)	233	237
Killing out %	54.4	52.2
Feed intakes, start to sale		
Silage DM (*kg*)	972	965
Compound (*kg*)	830	961

RM = Rosemaund EHF
CR = Crichton Royal Farm

Compared with the standard level of compound feeding of 2 kg/ day the relatively high levels of compound fed in these trials improved liveweight gains and shortened the period to sale. However these improvements were only marginal.

A later trial at Rosemaund showed that the cereal beef and silage beef production systems can be combined to finish cattle at around 450 kg at less than thirteen months of age (see Table 8.2). February born Holstein type bull calves were fed a 14 per cent CP barley-based compound to appetite from twelve weeks old. From eight months silage was on offer and the compound feeding was then gradually reduced to 2 kg/day over a four weeks period.

Table 8.2   Performance of bulls on a cereal beef/silage beef system

Number of bulls	20
Daily liveweight gain (12 weeks to sale) (*kg*)	1.02
Saleweight (*kg*)	450
Days to sale	393
Carcass weight (*kg*)	247
Killing out %	54.9
Feed intake, 12 weeks to sale	
Silage DM (*kg*)	850
Compound (*kg*)	1210

Source: Rosemaund EHF 1984–85.

This system combines two successful indoor methods and has the following attractions:

- It is particularly suitable for the finishing of Holstein type bulls.
- The silage requirement per head is around 3.4 tonnes compared with the 4.8 tonnes needed on the silage beef system. This reduction of almost 30 per cent is very welcome in areas where it is difficult to achieve big yields of high quality silage because of poor soils or adverse climatic conditions.
- Feeding silage during the summer months may be avoided. This removes the possibility of aerobic deterioration of the silage in mild summer weather. It also means that only one silo is needed whereas on the Rosemaund silage beef system at least two silos are normally required.
- The summer labour requirement is reduced because barley feeding requires less labour than silage feeding.
- Financial assessments show that compared with silage beef production the gross margin per head is less but the margin per hectare is similar.

Legume Silages

In the United Kingdom the principle forage legumes are white clover, red clover, lucerne and forage peas.

The clovers are more suited to the wetter climate of the west, whilst lucerne grows better in the drier eastern counties. To achieve efficient utilisation of red clover and lucerne these are best conserved, whilst white clover is normally grown and grazed in combination with grass.

Each of the legumes has a different crop duration and can be grown pure, or in combination with various companion crops. Basic relevant information is provided in Table 8.3.

Table 8.3   Sowing dates, duration and companion crops

	Lucerne	Red Clover	White Clover	Forage Peas
Time of sowing	Late March–early August	Late March–early August	Late March–early August	March–April
Suitable companion crop	Nil or Timothy/Meadow fescue	Nil or PRG	PRG	Spring barley
Crop duration	2–6 years	2–3 years	2–20 years	100 days

PRG = perennial ryegrass

Legumes offer the following advantages.

- Without the use of fertiliser nitrogen they offer a potential yield of up to 12 t DM/ha. However at Rosemaund EHF yields of lucerne and red clover have been in the range of 8–11 t DM/ha.

- Voluntary intake and efficiency of feed conversion are high.

- The digestibility of legumes, particularly of white clover, tends to decline at a slower rate than that of grasses as the crop matures. The penalty for delayed harvesting is therefore less.

- Legumes have a relatively high protein content compared with grasses and are a rich source of minerals.

They do have, however, several distinct drawbacks.

- They are low in soluble carbohydrates and have a poor buffering capacity. Therefore, when they are ensiled the use of a silage additive is normally advised to ensure an acceptable fermentation.

- Establishment can be difficult, and yields and persistency are variable.

- Legumes are generally lower in dry matter content than grass.

- Legumes have a high oestrogenic activity which, particularly if they are fed fresh, may adversely affect the fertility of breeding stock.

- Digestive problems, particularly bloat, can occur when legumes are fed fresh.

Conservation of legumes
With the exception of white clover the forage legumes perform best as cutting leys, and this also avoids the problems of bloat and poor utilisation which can occur when they are grazed. The making of legumes into hay in the United Kingdom is extremely difficult and high losses can occur as a result of poor preservation and from shattering of the leaf. Ensilage is very much to be preferred but requires special attention to achieve success. The following measures are considered important. First, wilt the crop to 20–25 per cent dry matter so that sugars are concentrated. Pure stands of low dry matter legumes such as red clover may require several days of wilting even in dry weather. Secondly, chop the herbage to facilitate the release of sugars and aid compaction of the crop in the silo. Thirdly apply an acid/formaldehyde additive—normally at double the rate required for grass.

The use of companion crops
Commonly legumes are grown with a companion grass crop. Mixtures usually attain a higher dry matter and a higher digestible organic matter yield. They have a higher soluble carbohydrate content and are better balanced nutritionally. Weed ingress is less of a problem with grass/clover mixtures and clover crowns are cushioned against wheel damage. However, lucerne is often grown pure and in these circumstances grass weeds can be more readily controlled chemically.

The feeding value of lucerne and red clover silages
Rosemaund EHF has compared pure red clover and pure lucerne silages with Italian ryegrass silage. These were fed to groups of steers in the late finishing period and the average results obtained over three years are shown in Table 8.4.

Table 8.4   Performance of cattle fed lucerne, red clover or Italian rye grass silages

	Italian ryegrass	Lucerne	Red Clover
Daily dry matter intake/head (*kg*)	7.80	9.20	8.70
Daily liveweight gain/head (*kg*)	0.85	0.73	0.78

Source: Rosemaund EHF.

Although intakes of the legume silages were higher, the higher energy content of the Italian ryegrass silage resulted in slightly better performance. The results in Table 8.4 were obtained when no supplementary compound was fed, and indicate that all these silages needed supplementing in order to achieve the target gain of one kg per day normal on indoor systems.

The animals fed the legume silages made the poorer liveweight gains and poorer carcass gains at Rosemaund and therefore had the greater need for supplementary feeding to achieve target weight gains. However, results from the Agricultural Research Institute of Northern Ireland and elsewhere showed similar carcass gains from legume silage and from ryegrass silage.

The feeding value of mixed grass/clover silages
White clover/grass silages have given equivalent or slightly better performance than grass silage when fed to beef cattle at both Rosemaund and Drayton EHFs and at Greenmount College in Northern Ireland. Performance from red clover/grass silages at the latter institute has not been as good as that from grass silage.

Output per hectare of forage
Despite the ability of legume silages, whether pure or as grass mixtures, to produce liveweight gains similar to those from pure grass silage, their output of carcass meat per hectare is considerably lower than that from grass silage. There are two reasons for this. First, killing-out or dressing percentage from animals fed legume silage is reduced. Secondly, the yield of crop dry matter is significantly lower (and more variable) than that of pure grass receiving heavy dressings of nitrogenous fertiliser. This is particularly so when the reduced yield in their year of establishment is taken into account.

Therefore a combination of lower yields and higher intakes can reduce stocking rates where legume silages are fed.

Estimates from the EHFs and from Northern Ireland indicate the relative outputs shown in Table 8.5 from various leys grown under optimum conditions.

Table 8.5 **Relative output of carcass meat per hectare**

Silage crop	*Relative output (grass = 100)*
Grass + 300–400 kg N/ha/annum	100
Grass/white clover + 50 kg N/ha/annum	75
Grass/red clover + 50 kg N/ha/annum	70–75
Red clover, nil N	70–75
Lucerne, nil N	70

To counterbalance these lower outputs, leguminous leys can give a saving of 250–400 kg of nitrogen/ha (depending on location). However, their phosphate and potash requirements are somewhat higher than those of a grass ley.

At current costs the saving of expenditure on nitrogenous fertiliser, resulting from the growing of leguminous leys, does not balance the loss resulting from the lower meat output per hectare. However, higher nitrogen costs in the future and environmental considerations may make the nitrogen fixing legumes both financially competitive and environmentally attractive, as would an increase in demand for organically produced food.

Forage pea silage

The forage pea is an annual crop which reaches the silage cutting stage around fourteen weeks after sowing. Peas are often used as a cover crop for grassland maiden seeds and for young legumes such as lucerne or red clover. They are best sown with spring barley or with an erect variety of Italian ryegrass. These companion crops provide support for the peas and help to prevent lodging onto the undersown seeds. Lodging can be a problem when peas are sown alone.

Yield of forage peas

Yields are variable. ADAS and the Scottish Colleges' trials have shown a range from 3 to 12 t DM/ha. The occurrence of low yields was attributed to poor establishment, weed infestation and lodging. Birds can inflict serious damage at sowing time.

Typical yields on well drained adequate moisture status soils are given in the Table 8.6.

Table 8.6 Yields of forage peas cut fourteen weeks after sowing (t/ha)

	Peas alone	Peas and barley
Fresh yield	30	40
Dry matter yield	6	8

Forage pea silage quality

In common with all legumes, forage peas are low in soluble sugars. They should be wilted and chopped and receive an acid/formalin additive to promote a desirable fermentation. This objective is also favoured by sowing peas along with cereals or ryegrass. However, if cereals form a high proportion of the mix, overheating in the silo can be a problem. For this reason mixtures should be well chopped and consolidated, and used quickly during the feeding period.

Table 8.7   Typical analyses of forage pea silages

	Peas only	Peas + barley	Peas + Italian ryegrass	Italian ryegrass (for comparison)
Dry Matter (%)	23.0	27.0	24.0	25.0
pH	4.3	4.1	4.2	4.0
ME (MJ/kg DM)	9.5	9.4	10.0	10.6
CP (g/kg DM)	190	130	170	160

Source: Derived from ADAS and Scottish Colleges' information.

An analysis of forage pea silages is detailed in Table 8.7. The high protein levels and low energy (ME) values of peas are typical of all legumes.

Experience has shown that animal intake of pure forage pea silage and of mixed silages containing peas is somewhat higher than that of grass silage and liveweight gains are therefore not likely to suffer greatly as a result of their lower energy values.

In summary, the forage pea crop is a useful annual legume of high protein content. However, its main value (particularly when grown in combination with cereals) is to act as a high yielding nurse crop for undersown grasses and legumes.

Whole Crop Cereal Silage

All cereal crops can be forage harvested at the dough ripe stage of grain maturity and successfully ensiled, but spring barley is the most frequent choice, giving a typical yield of around 8–10 t of DM/ha, although yields of up to 15 t/ha have been reported.

Advantages of whole crop cereal silage are:

• It does not necessitate multiple cutting.
• It requires little or no field wilting.
• A favourable fermentation in the silo is normally assured.
• In-field losses at harvest are low.
• It forms an ideal cover crop for establishing grasses and/or legumes.

There are, however, some draw-backs:

• It is not a break crop in an arable rotation.
• Yields are normally lower than those of grass leys.
• It is particularly subject to deterioration at the silage face during the feeding period.
• Cereal silage has lower energy and protein contents than those of high quality grass silage.

Whole crop cereal silage quality
The feeding value (in the dry matter) of whole crop cereal silage is similar to that of average hay but below that of average grass silage. A typical analysis of whole crop cereal silage is given in Table 8.8.

Table 8.8   Typical analysis of whole crop cereal silage

Dry matter (%)	35–60
ME (*MJ/kg DM*)	9.5
CP (*g/kg DM*)	95

Maximum yields of whole crop cereal silage are obtained when the crop is ensiled at a dry matter content of 50 to 60 per cent. Undersown whole crops are normally best harvested at a dry matter content of approximately 40 per cent.

Digestibility declines as the crop matures. Unlike maize the fall in the quality of the stem is not matched by an increasing contribution from the grain.

Because whole crop cereals are normally high in both dry matter and sugar contents, their fermentation is normally satisfactory, but care must be taken to avoid aerobic spoilage and moulding when feeding. It is therefore important to chop the crop, consolidate well, and use the silage quickly during feeding.

Animal performance
Based on ADAS experience, cattle fed whole crop cereal silage are likely to give poorer liveweight gains than those fed good quality grass silage, and this is to be expected in view of the lower feed value indicated on analysis. An additional 2 kg/head/day of compound feed may be needed in conjunction with whole crop cereal silage to achieve target gains. However, trials conducted at the North of Scotland College and at the Grassland Research Institute have shown that treatment of the whole crop cereal with urea, ammonia, or sodium hydroxide will considerably enhance its feed value.

Summary
Whole crop cereal silage is unlikely to equal high quality grass silage in terms of yield, feeding quality, or resultant animal performance. It does, however, have an important place as a cover crop for young grass and legume seeds. Treatment with alkaline additives shows exciting potential.

Roots

All roots should be free from excessive soil contamination as this may cause digestive upsets and reduce intake.

Roots fed to young stock below 250 kg liveweight should be chopped or pulped to increase consumption. However, the feeding of whole roots to older cattle is not detrimental to intake or performance once the beasts are fully accustomed to this diet. Chopping does, however, prevent cattle from wasting roots (especially swedes and beet) when they remove them from the troughs.

Processing may also be beneficial during the period when cattle are changing their calf teeth, and may also minimise the risk of choking on some roots, especially potatoes and carrots. Very high dry matter roots, such as certain varieties of fodder beet, are quite hard and are more likely to need pulping or slicing to encourage consumption.

The root crops available for feeding indoors to beef cattle can be conveniently divided into two categories. First, roots intended for human consumption but surplus to requirements, such as potatoes, carrots, parsnips, sugar beet and beet root. Secondly, roots grown specifically for animal feed—fodder beet, mangels, turnips and swedes.

Market Surpluses

Occasionally surplus roots are available for animal feeding. Typical nutritive values of those most commonly available are shown in Table 8.9.

Table 8.9   Nutritive value of surplus root crops

Root	Dry Matter %	In the dry matter ME (MJ/kg)	DCP (g/kg)	Kg required to replace 1 kg barley
Potatoes	20	12.5	45	4.5
Carrots	11	12.8	60	8.0
Parsnips	16	13.0	70	5.5
Sugar beet	20	13.7	35	4.0
Beetroot	13	12.1	90	7.0
Barley (for comparison)	86	12.9	80	1.0

ME = Metabolisable Energy.
DCP = Digestible Crude Protein.
Source: ADAS Advisory Leaflet 820.

All have an ME (on a DM basis) similar to that of barley. The majority, however, are low in protein. In formulating rations containing these roots, additional protein (normally of good quality, e.g. soya bean or fishmeal) will be required along with a suitable balance of minerals and vitamins. For safe ration formulation, the maximum quantity fed should be limited to the equivalent of 1 kg of barley per 100 kg animal liveweight.

Potatoes

These are higher in dry matter than the majority of the other root crops, but lower in digestible crude protein. On average, about 4 to 5 kg of potatoes will replace 1 kg of barley to give an equivalent energy intake, but protein needs are not likely to be met. Thus for a 400 kg beast, a practical diet would contain no more than 20 kg/head/day of potatoes. In trials at the Rowett Research Institute and Gleadthorpe EHF, however, larger quantities have been fed ad lib to cattle of over 250 kg liveweight with no adverse effects. The results are shown in Table 8.10.

Table 8.10   Replacement of barley by potatoes

Ration Composition (on DM basis)	100% Barley*	34% Barley* 66% Potatoes	13% Barley* 87% Potatoes
Daily liveweight gain (kg)	1.0	1.2	1.1
Dry matter intake (kg)	5.9	6.4	6.0
Average fresh intake (kg)			
Barley*	6.9	2.5	0.9
Potatoes	—	22.0	27.0

* With added protein plus minerals and vitamins
Source: Rowett Research Institute (Kay et al 1972).

Gleadthorpe EHF has finished 400 kg steers on 32 kg/day of potatoes with 1 kg barley and 0.5 kg fishmeal/day plus 2.5 kg straw and minerals/vitamins.

There are two possible dangers from feeding potatoes to beef cattle. Choking is frequently cited as a potential hazard. To avoid this potatoes can be fed sliced or pulped. Alternatively, whole potatoes can be given but they should be fed on the ground with a suitable feed barrier to prevent cattle raising their heads above shoulder height. Poisoning from potato sprouts which contain solanines may also occur. The sprouts must be removed prior to feeding.

Carrots

Normally 8 kg of carrots will replace 1 kg of barley, but protein content is slightly lower. Their orange colour is due to beta carotene (a precursor of Vitamin A) and feeding high levels in the diet can lead to a yellow carcass fat. Instances of vitamin A toxicity have been reported where very large quantities have been fed. Choking can be a problem when certain sizes of carrots are fed.

Parsnips
They are higher in dry matter than carrots and the carotene content is lower but still significant. About 5.5 kg parsnips will replace 1 kg barley.

Sugar beet
These are similar to fodder beet (see page 112), but tend to be higher in dry matter and very low in protein.

Beetroot
Contains rather more protein than other beets but is lower in dry matter. The urine of beetroot fed cattle may have a red colouration. From 7 to 8 kg of beetroot will replace 1 kg barley.

Roots Grown for Animal Feed
The most important are swedes and fodder beet. Their biggest advantage is their ability to produce very high yields of dry matter and metabolisable energy (ME) per hectare as demonstrated in Table 8.11.

Table 8.11   Typical yields of swedes, fodder beet and barley

		Dry Matter (t/ha)	ME (GJ/ha)	DCP (Kg/ha)
Swedes	(roots)	6	83	390
	(tops)	2	21	260
Fodder beet	(roots)	11	130	550
	(tops)	3	30	270
Barley	(grain)	5	65	400
	(straw)	2	14	20

GJ/ha = Gigajoules/ha.
The yields shown above are typical in lowland areas in England and Wales.
Source: ADAS.

Swedes
Swedes are similar to barley in terms of energy and protein (see Table 8.12). They are very low in dry matter although this varies with variety. Purple skinned varieties have the lowest dry matter (less than 10 per cent) and green skinned varieties are of higher dry matter content.

Table 8.12 Relative nutritive value of swedes, fodder beet and barley

	Dry Matter %	In the dry matter ME (MJ/kg)	DCP (g/kg)	Kg required to replace 1 kg barley
Swedes	10	12.8	90	8.5
Fodder Beet	18	12.0	50	5.0
Barley (for comparison)	86	12.9	80	1.0

Source: ADAS.

From 8 to 9 kg of swedes will normally replace 1 kg of barley. In most commercial rations the weight of swedes fed should not exceed 7–8 kg per 100 kg liveweight. However, as with potatoes, considerably more have occasionally been given as a replacement for barley with no ill effects and only a small reduction in animal performance (see Table 8.13).

Table 8.13 Replacement of barley by swedes

Ration composition (on DM basis)	100% Barley*	50% Barley* 50% Swedes	100% Swedes*
Daily liveweight gain (kg)	1.0	1.0	0.9
Daily dry matter intake (kg)	6.2	6.3	6.0
Average daily fresh intake (kg):			
Barley*	7.3	3.7	—
Swedes*	—	33.0	62.0

* With added minerals and vitamins plus protein as necessary
Source: Rowett Research Institute (Kay et al 1972).

Similar results have been achieved at Rosemaund EHF where swedes were fed ad lib.

In practice, the high intakes which occur on ad lib swede feeding lead to excessive urine output and unacceptably high bedding straw usage.

The protein supplementation requirements of diets containing swedes are similar to those containing barley and a suitable mineral/vitamin supplement should be provided.

Fodder beet
The ME of fodder beet is around 90 per cent of that of barley, but the protein level is considerably lower (see Table 8.12). The dry matter varies from around 14 per cent up to 22 per cent depending on variety. In general, the deeper the root grows in the soil, the higher the dry matter percentage. The keeping quality of the medium

Mechanised feeding of whole fodder beet to Friesian bulls. Photograph Gleadthorpe EHF

and high dry matter varieties is considerably better than that of swedes and the roots can often be fed until the end of May.

On average, 5 kg of high dry matter roots will replace 1 kg of barley in terms of energy. Up to 40 kg/head/day of fresh beet have been fed to finishing cattle at both Gleadthorpe and Rosemaund EHFs. Again, urine output is quite high and ample straw bedding is required.

Suitable rations (per beast per day) fed on the EHFs are:

High beet content ration—for 400 kg beast; daily gain 1.0 kg
40 kg beet (16 per cent dry matter)
0.8 kg soya bean meal
0.5 kg barley straw

Beet and barley—for 400 kg beast; daily gain 1.2 kg
34 kg beet (16 per cent dry matter)
2.7 kg rolled barley
0.7 kg soya bean meal
0.5 kg straw

Beet and grass silage—for 400 kg beast; daily gain 0.9 kg
10 kg beet (16 per cent dry matter)
24 kg silage (25 per cent dry matter; 10.3 ME)
0.4 kg soya bean meal

Supplementation of fodder beet rations
Additional protein of high quality is necessary in all rations containing fodder beet even when sufficient crude protein is present (e.g. in silage-based rations). Urea is an unsuitable protein supplement. Fishmeal appears to give best results but soya bean meal should be a perfectly adequate protein supplement.

Compared with barley fodder beet is high in calcium and low in phosphorus. A suitable supplement should contain 14 per cent of calcium and 9 per cent of phosphorus with 300,000 i.u. vitamin A and 600–800 i.u. vitamin E per kg of supplement.

A System of Indoor Fodder Beet Feeding

Gleadthorpe EHF has developed an indoor finishing system for spring-born calves. It relies mainly on whole fodder beet feeding from late September onwards, aiming to finish cattle the following spring at twelve to fourteen months of age. See Table 8.14.

Table 8.14   Input and output targets for the Friesian bull

	Liveweight (kg)
At arrival in April	50
Early October at start of beet feeding	250
Slaughter—April/May (full fed)	460
Daily gain October to slaughter	1.1
Feed consumed from twelve weeks (kg):	
Compound	460
Fishmeal	115
Fodder beet (18% DM) (*t*)	6.8

Source: Gleadthorpe EHF.

Feeding differs in the rearing and the finishing periods.

Rearing period—arrival to early October: calves early weaned and fed barley straw ad lib, plus up to 4 kg/head/day of 14 per cent crude protein barley mix.

Finishing period—early October to slaughter in April–May: whole fodder beet are given ad lib with an average of 0.6 kg of fishmeal or 0.9 kg of soya bean meal daily with a mineral/vitamin supplement. Barley straw is always available.

The diet change takes place over a two to three weeks period in early October, and must be gradual, to allow cattle to adjust to the higher sugar content of their diet. This adjustment is promoted by the initial feeding of chopped (rather than whole) beet.

Daily consumption of fodder beet rises from 18 kg in October to 30 kg in February and ultimately approaches 40 kg. Straw consumption is negligible at 0.5 kg/head/day, but the bedding straw requirement is high.

Hereford *x* Friesians have also been finished at Gleadthorpe EHF on this system as shown in Table 8.15.

Table 8.15 Targets for Hereford x Friesians finished on fodder beet (kg)

	Heifer	*Steer*	*Bull*
Weight in October at start of beet feeding	190	220	250
Weight at slaughter (full fed)	360	430	445
Daily gain	0.9	1.0	1.1
Feed consumed from 12 weeks:			
Compound	360	400	450
Fishmeal	80	80	100
Fodder beet (18% DM) (*t*)	4.5	5.5	6.3

Source: Derived from information at Gleadthorpe EHF.

Hereford *x* Friesian steers (and bulls in particular) are suitable for this finishing system but heifers can become over-fat at very low liveweights. For this reason, intakes of beet should be restricted for heifers during the final two months of finishing.

The Feeding of Root-crop Tops

Potato haulm is poisonous, but all the other root crops mentioned above produce significant yields of tops which have potential feed value. It can be difficult and costly to utilise tops by feeding them to stock, and for this reason many are ploughed in.

If tops are to be used for feeding it is important to avoid soil contamination, and with the exception of beet tops, which should be pre-wilted, they are best fed green. However, clean tops can be ensiled although this frequently results in heavy in-silo losses.

Brassica tops, if fed in large quantities, may give rise to anaemia and unthriftiness. This is due to the presence in the tops of small amounts of S-methyl cysteine sulphoxide (SMCO). The feeding of beet tops can cause scouring if they are not wilted and introduced into the diet gradually.

Straw Feeding

Straw is a cheap source of energy. Unfortunately, compared with hay, straw contains a lower proportion of energy, protein, minerals and vitamins. Moreover it is very variable in nutritional value, this variability existing between the cereal crops and between individual varieties. Straws from modern varieties of autumn sown cereals are generally not significantly poorer in feed value than those of spring sown varieties, but wheat straws are poorer than barley straws. See Table 8.16 for the average composition of untreated straws.

Table 8.16   Average composition of untreated straws

Straw type	DM %	CP %	'D' value	ME (MJ/kg DM)	DCP (g/kg DM)
Barley	85	3	45	6.7	15
Wheat	85	4	42	6.0	17
Oats	85	3	50	7.2	11

Source: ADAS.

The ADAS Feed Evaluation Unit has found that most untreated straws have a 'D' value within the range thirty to fifty. Therefore untreated straw, fed alone, is incapable of maintaining any beef animal and even when supplemented it has no place in the high liveweight gain systems described in this book.

The feed value of straw may be improved by mechanical and by chemical means.

Mechanical processing
Mechanical processing by chopping or grinding improves feed intakes because rumen retention time is reduced. However, it has no effect on the digestibility of the straw. Chopping is expensive, but essential if straw is to be mixed with other feeds, and fine grinding allows straw to be incorporated in cubed rations.

Chemical processing
Sodium hydroxide (caustic soda) or ammonia may be used to break down the outer lignin layer of straw in order to make available usable energy. A wide range of machinery is available to chop straw and to treat it with caustic soda. However this treatment does result in increased production of urine, due to the high sodium content of treated straw, and bedding requirements are consequently increased.

Ammonia may be applied in the anhydrous form to bales in an insulated air-tight oven, or it may be injected (often as aqueous ammonia) into a plastic covered, sealed stack of straw bales. The ammonia treatment improves the digestibility of straw as does treatment with caustic soda. However it has the advantage that the straw remains in the long form and can be fed in racks. It may also be of superior palatability and will show an increased protein content compared with straw treated with sodium hydroxide.

Effects of straw treatment

Table 8.17   Effect of alkali treatment on in vivo 'D' value

Chemical	Straw	Increase in 'D' value (percentage units)
Sodium hydroxide	Barley	15
	Wheat	15
Ammonia	Barley	9
	Wheat	10

Source: MAFF leaflet 618.

Table 8.17 shows by what percentage the 'D' value of straw is increased by treatment with either caustic soda or ammonia. Treatment also brings the energy content of straw up to around that of medium quality hay, as shown in Table 8.18.

Table 8.18   Feeding value of hay and of treated straws

	Average hay	Sodium hydroxide treated		Ammonia treated	
		Barley straw	Wheat straw	Barley straw	Wheat straw
'D' Value	56	64	57	50	44
ME (MJ/kg DM)	8.3	9.0	7.7	7.2	7.2
CP % in DM	8.7	4.3	4.0	6.7	6.4

Source: ADAS and MAFF Leaflet 618.

Use of straw in beef rations
Treated straw may be regarded as a medium quality bulk feed suitable for use (suitably supplemented) in beef enterprises where high liveweight gains are not required. These could include store cattle, suckler cows and rearing heifers.

Attempts to utilise treated straw as the main bulk feed for finishing cattle have met with little success. In Ireland, ammoniated straw was fed to mature steers and supplemented with 2 kg/head/day of a barley/soya mix and the gain recorded was 0.59 kg/day. At Rosemaund, heifers fed treated straw and a barley/white fishmeal compound at 3 kg/day gained at 0.84 kg/day. The cost of the supplementary feed needed to achieve target gains is prohibitive. However, in the cereal beef unit a supply of fibre is advisable to prevent bloat and other digestive disorders. The inclusion of no more than 10 per cent tub-

ground treated straw in the cubes on offer ensures that the animals ingest the necessary fibre. An alternative is to offer long straw in racks. As mentioned on page 114 there is also a place for straw feeding where the ad lib feeding of roots is practised.

There is no reason to feed treated straw in the silage beef unit since adequate fibre is available in the silage.

Silage Effluent as a Feed

Silage effluent is potentially an extremely potent pollutant of the environment. If it enters watercourses dissolved oxygen levels are reduced and fish and other aquatic life may die. The Biochemical Oxygen Demand (BOD) is a measure of the strength of an effluent, and the BOD of silage effluent is some 150 times greater than that of raw domestic sewage. Effluents can be discharged into watercourses only with the written consent of the Water Authority, and contravention of the Control of Pollution Act 1974 may lead to prosecution.

Volume of effluent
This is related to the moisture content of the forage ensiled as shown in Table 8.19.

Table 8.19   Expected silage effluent from grass in a
clamp silo

Grass dry matter (%)	Effluent produced (litres per tonne)
15	150–330
20	60–220
25	10–110

Source: MAFF booklet 2429.

It can be seen that a silo containing 500 tonnes of 20 per cent DM grass can give up to 110,000 litres (24,000 gallons) of effluent.

Feed value of effluent
This depends largely on the dry matter content of the effluent which has averaged 6 per cent with a range of 4 per cent to 12 per cent. The energy content in the DM is quite high (similar to that of barley) and the crude protein content is high at an average of 22 per cent. Much of this crude protein is, however, degradable in the rumen.

Scandinavian experience suggests that effluent intake should be restricted to 5 litres per 100 kg liveweight. Thus a beef animal weighing 300 kg should be allowed no more than 15 litres (3 gallons) per day.

Volumes of almost double this have been fed in the North of Scotland with no ill effects.

On-farm experience
Effluent is normally collected in a sump before pumping to the storage tank from which it is piped (frequently under gravity) to the cattle. If the storage period exceeds four days a preservative must be added. Formalin added at 3 litres per 1,000 litres of effluent has ensured preservation for up to one year. Most users have reported good palatability with cattle drinking effluent in preference to water.

In trials, including several at Hillsborough Research Station in Northern Ireland, the intake of effluent by beef cattle has been very variable, but has averaged around 14 litres per day. This has replaced 0.75 kg of compound without loss of performance.

Three new approaches to the collection and feeding of silage effluent have been developed recently. First the addition to grass at ensilage of ground cereals (Welsh Plant Breeding Station) or feeds based on sugar beet pulp. By these means effluent flow may be reduced and the energy content of the silage enhanced. Secondly, the layering of chopped straw in the silo at the time of loading with grass. Liscombe EHF has reported that this technique may slow the ensilage operation and it will inevitably reduce the silage feed value. Thirdly, the development at Liscombe EHF of a straw bale barrier as a 'lining' of the silage clamp floor. This has reduced effluent flow by around 30 per cent when 18–20 per cent DM grass was ensiled, but the resultant effluent-soaked bales are not of high feed value.

Summary
The results quoted above indicate that the feeding of effluent to beef cattle can give a saving in compound feed costs of about 10p/beast/day. It may also be highly beneficial in reducing the incidence of pollution. However, storage facilities are necessary to ensure the continuous availability of effluent, and the cost of these must not outweigh the feed saving benefit.

Chapter 9

FINANCIAL CONSIDERATIONS

THIS CHAPTER gives specimen layouts which should enable the reader to calculate gross margins for the cereal, grass silage, maize silage and fodder beet beef systems. The quantities of inputs and outputs (e.g. compound use and saleweights) are typical of those attainable in well managed units. However, these levels could vary a great deal in practice depending on individual managerial skill.

Remember, *a gross margin is not a profit margin*. Fixed or overhead costs including labour, rent, machinery and interest charges on borrowed capital have to be deducted from gross margins to arrive at a profit margin.

These fixed cost items will vary according to individual farm circumstances and the particular beef system adopted. Great care is needed, therefore, in comparing gross margins of different beef systems. The full impact of each system on farm overhead costs must be fully considered before arriving at any decision on the most appropriate beef system for a particular farm.

It is important to remember that improved profitability is not simply a matter of substituting a high output indoor system for a low output grazing system. Each system has very different requirements for basic farm resources, particularly labour, machinery, buildings and working capital. All these factors need to be taken into account at the planning stage.

Gross Margin Analyses

**Table 9.1   Cereal beef system. Gross Margin calculation
Friesian bulls**

Output:
Sale of beast
        420 kg liveweight (235 kg carcass)   @ ...... p/kg   ....
Less cost of calf (including 6% mortality)   ....
                              Total output   £

Variable costs:

To 12	Milk substitute	13 kg @ ...... £/t	....
weeks	Early weaning compound	160 kg @ ...... £/t	....
	14% CP Compound	1700 kg @ ...... £/t	....
	Straw (bedding)	900 kg @ ...... £/t	....
	Vet and medicine		....
	Sundries (marketing costs etc)		....
		Total variable costs	£
		Gross Margin/head	£

**Table 9.2   Cereal beef system. Gross Margin calculation
Charolais x Friesian bulls**

Output:

Sale of beast

470 kg liveweight (270 kg carcass)	@ ...... p/kg	....
Less cost of calf (including 6% mortality)		....
	Total output	£

Variable costs:

To 12	Milk substitute	13 kg @ ...... £/t	....
weeks	Early weaning compound	160 kg @ ...... £/t	....
	14% CP Compound	1720 kg @ ...... £/t	....
	Straw (bedding)	900 kg @ ...... £/t	....
	Vet and medicine		....
	Sundries (marketing costs etc)		....
		Total variable costs	£
		Gross Margin/head	£

Comments on financial performance of cereal beef systems

The average gross margin/head in MLC recorded units in 1984 was £72 for units rearing their calves and £56 for those using reared calves. Top third producers achieved £95 and £86 respectively; better feed conversion efficiency was the major reason for their better performance with higher slaughter weights and higher daily gains being achieved. In eleven units finishing Continental cross calves, the average gross margin in 1984 was £48, with a range from £6 to £114, this despite faster growth rates, heavier slaughter weights and approximately 10 per cent premium per kg liveweight. These results show that high standards of management and astute marketing are vital if the higher investment in Continental cross calves is to pay off. In general, cereal beef profitability is low. It not only demands the highest levels of management for success, but also meat buyers who are prepared to pay a premium for young, well fleshed carcasses with white (as opposed to cream) fat colour. However, this system can release land for other purposes and this opportunity use can be valuable.

Table 9.3   Grass silage beef system. Gross Margin calculation
Hereford x Friesian bulls

Output:
Sale of beast
          425 kg liveweight (230 kg carcass)   @ ...... p/kg   ....
Less cost of calf (including 6% mortality)   ....
                                              Total output   £
                                                                                 ‾‾‾‾‾‾‾‾‾‾

Variable costs:
To 12   Milk substitute   13 kg @ ...... £/t   ....
weeks   Early weaner compound   120 kg @ ...... £/t   ....
         Home mixed compound   600 kg @ ...... £/t   ....
         Silage   1.2 t DM (4.8 t fresh) @ ...... £/t   ....
         Straw (bedding)   1300 kg @ ...... £/t   ....
         Vet and medicine   ....
         Sundries (marketing costs etc)   ....
                                              Total variable costs   £
                                                                            ‾‾‾‾‾‾‾‾‾‾
                                              Gross Margin/head   £
                                                                            ‾‾‾‾‾‾‾‾‾‾
         Stocking rate: 8 beasts/hectare*
         Gross Margin/hectare   £
                                                                            ‾‾‾‾‾‾‾‾‾‾

* Based on silage yield of 10 t DM/ha

Table 9.4   Grass silage beef system. Gross Margin calculation
Friesian bulls (normal saleweight)

Output:
Sale of beast
          450 kg liveweight (240 kg carcass)   @ ...... p/kg   ....
Less cost of calf (including 6% mortality)   ....
                                              Total output   £
                                                                                 ‾‾‾‾‾‾‾‾‾‾

Variable costs:
To 12   Milk substitute   13 kg @ ...... £/t   ....
weeks   Early weaner compound   120 kg @ ...... £/t   ....
         Home mixed compound   770 kg @ ...... £/t   ....
         Silage   1.4 t DM (5.6 t fresh) @ ...... £/t   ....
         Straw (bedding)   1350 kg @ ...... £/t   ....
         Vet and medicine   ....
         Sundries (marketing costs etc)   ....
                                              Total variable costs   £
                                                                            ‾‾‾‾‾‾‾‾‾‾

                        Gross Margin/head   £
                                                 _____

Stocking rate: 7 beasts/hectare*
Gross Margin/hectare   £
                                                 _____

 * Based on silage yield of 10 t DM/ha

   Rising calf prices and changing market preferences are causing
many silage beef operators to consider taking cattle to heavier weights
at sale. Friesian and Continental cross bulls are suitable for this
purpose, and appropriate input and output figures for these and for
Hereford x Friesian heifers are given in Table 9.5. The provision of
these figures and the use of the format in Tables 9.1 to 9.4 will
enable the prospective gross margins per head and per hectare to be
calculated.

Table 9.5   Silage beef system. Gross Margin calculations

	Friesian bulls (heavy saleweight)	Charolais x Friesian bulls	Hereford x Friesian heifers
Sale liveweight (carcass weight) (kg)	520 (290)	550 (315)	360 (195)
Home mixed compound (kg)	920	930	400
Silage (t DM)	1.8	1.8	1.0
Straw for bedding (kg)	1500	1500	1300
Stocking rate. (beasts/ha*)	5.5	5.5	10.0

* Based on silage yield of 10 t DM/ha

Comments on financial performance of grass silage beef
The average gross margin for 1984 from sixty-two groups of reared
calves on MLC recorded units was £122 per head and £817 per hectare.
Top third producers achieved margins of £153 per head and £1,220
per hectare. Higher daily gains and lower compound consumption
coupled with higher stocking rates (plus 1.3 beasts/ha) were responsible for top third producer success. The main reason for the poor
performance of the bottom third producers was high compound
consumption. In most cases this resulted from the need to feed extra
cereals with poorer quality silage.
   Hereford x Friesian bulls achieved similar gross margins per head
to Friesian bulls; the higher price for the Hereford x Friesian calf was
cancelled out by lower compound input costs. Total returns were
similar for both breed types because the lower saleweight of the
Hereford cross was compensated by a higher price per kg at sale. This
result contrasts with that from Rosemaund EHF where Friesian bulls
have given better margins than Hereford crosses. The high quality

silage made at Rosemaund has always ensured low compound inputs to Friesian bulls.

The gross margin per hectare from MLC recorded Hereford x Friesians exceeded that from Friesian bulls by £44 per hectare because of their lower silage consumption and consequent higher stocking rate.

Gross margins from Friesian steers were some 25 per cent to 30 per cent lower than for Friesian bulls.

To achieve worthwhile margins from silage beef, heavy yields of high quality silage must be produced so that compound costs can be minimised and high stocking rates attained.

Table 9.6   Maize silage beef system. Gross Margin calculation
Friesian bulls

Output:
Sale of beast
             465 kg liveweight (255 kg carcass)   @ ...... p/kg   ....
Less cost of calf (including 6% mortality)   ....
                                                      Total output   £
                                                                          _____

Variable costs:

To 12 weeks			
Milk substitute	13 kg @ ...... £/t	....
Early weaner compound	120 kg @ ...... £/t	....
35% CP Compound	500 kg @ ...... £/t	....
Barley	150 kg @ ...... £/t	....
Silage	1.75 t DM (6.25 t fresh) @ ...... £/t	....
Straw (bedding)	1300 kg @ ...... £/t	....
Vet and medicine		....
Sundries (marketing costs etc)		....

                                       Total variable costs   £
                                                                 _____

                                       Gross Margin/head   £
                                                                 _____

Stocking rate: 5.7 beasts/hectare*
Gross Margin/hectare   £
                                                                 _____

* Based on silage yield of 10 t DM/ha

Comments on the financial performance of maize silage beef systems
There are too few MLC recorded units to allow a full financial analysis but gross margins of around £1000/ha were being achieved in Southern England in the 1980s. The most crucial factor determining margin will obviously be the yield of forage maize per hectare.

Comments on financial performance of legume silage beef
To the authors' knowledge there are currently no producers feeding all-legume silages as the sole bulk feed. However, it can be safely assumed that the resultant gross margins would be considerably lower than from grass or grass/legume silages. Current evidence indicates that silage yields from all-legume leys are lower and animal intakes are higher.

Large increases in the cost of nitrogenous fertiliser in the future could increase the attraction of leguminous leys. Gross margins, however, will be very dependent on the yields that can be achieved from the legumes. The most promising when grown at Rosemaund have been lucerne and red clover, but only under ideal growing conditions have they yielded reliably at around 10 t DM/ha.

Table 9.7   Fodder beet beef. Gross Margin calculation
Friesian bulls

Output:
Sale of beast

	450 kg liveweight (245 kg carcass)	@ ...... p/kg	....
Less cost of calf (including 6% mortality)			....
		Total output £	_____

Variable costs:

To 12	Milk substitute	13 kg @ ...... £/t	....
weeks	Early weaner compound	120 kg @ ...... £/t	....
	Compound	460 kg @ ...... £/t	....
	Fishmeal	115 kg @ ...... £/t	....
	Fodder beet	1.2 t DM (6.8 t fresh) @ ...... £/t	....
	Straw (bedding)	1500 kg @ ...... £/t	....
	Vet and medicine		....
	Sundries (marketing costs etc)		....
		Total variable costs £	_____
		Gross Margin/head £	_____

Stocking rate: 10 beasts/hectare*
Gross Margin/hectare   £   _____

* Based on yield of 12 t DM/ha

Comments on financial performance of beef from fodder beet
Only Gleadthorpe EHF has published financial returns from this system. In 1982/83 and 1983/84, gross margins per hectare ranged from £1,600 to £1,800 for both heifers and steers stocked at 13–14 beasts/ha of fodder beet. In 1984/85, Friesian bulls stocked at between

8 and 10 beasts/ha gave gross margins of between £1,500 and £2,000/ha.

In commercial practice, margins are likely to be similar to the above provided adequate yields of fodder beet can be grown. Gleadthorpe's yield of 12 t DM/ha without irrigation should be easily within the reach of growers on suitable soil types, particularly if they have had previous experience of growing sugar beet.

Fixed Costs Implications

Fixed costs are defined by John Nix of Wye College (University of London) as those costs which are unallocated in determining enterprise gross margins. He tabulates them as follows:
>   Regular labour (paid)
>   Regular labour (unpaid)
>   Machinery depreciation
>   Machinery repairs
>   Fuel and oil
>   Unallocated contract charges
>   Vehicle tax and insurance
>   Rent (or rental value) and rates
>   General overhead expenses

Fixed costs alter more radically than variable costs when major enterprise changes occur. If enterprise substitutions are of a minor nature fixed costs may alter little from year to year.

In 1986, Nix estimates them to vary from £850/ha for mainly dairying farms, to £360/ha for beef/sheep holdings. Fixed costs are not given for indoor beef but levels approaching those of dairying are more likely than beef/sheep costs.

The costs of buildings, machinery and labour and interest charges are the main items which are subject to change with switches of beef production system or the introduction of a new enterprise.

Changing to cereal beef is likely to have the least influence on fixed costs. Machinery costs are comparatively low, requiring minimally a small tractor and trailer to move feed and bedding straw, up to an investment in mill and mix equipment and automated feed dispensers on very large units. Labour costs are also likely to be relatively low because of the dry easy to handle feed.

All the other systems outlined are based on bulky high moisture content feed. Consequently the fixed costs for food storage and handling are likely to be relatively high. Of the bulky feed systems, those based on silage are likely to incur the highest costs for food storage and for machinery and labour for the making and feeding of the silage. The cost of silos is particularly important here and this requirement is discussed in Chapter 10. Systems based on roots should carry a rather lower food storage and machinery cost.

Working Capital Implications

The working capital requirement is that needed to purchase calves and compound feed, and to cover the other variable costs. It can be defined in two ways.

The average annual working capital requirement per head (and per hectare on forage based systems).
The peak working capital, i.e. the peak reached before first sales.

Average working capital/beast = calf cost (£) + half variable cost (£)
Interest cost/beast = Working capital (£) × interest rate (%) × time on farm (months) ÷ 12
Interest cost/ha = interest cost/beast (£) × stocking rate (beasts/ha)

The working capital profiles for indoor systems fall conveniently into two categories.

- Intensive cereal beef which provides sales within ten to twelve months from birth of the calf.
- Forage (silage or roots) based systems which provide sales after twelve to fifteen months from birth.

Their capital profiles are represented diagramatically in Figures 9.1(a) and (b).

Assuming batch buying of calves to allow sales throughout the year and starting with ten day old calves the annual average working capital is currently similar for the two types of system. Their peak requirements differ significantly however in both level and timing. In the twelve to fifteen month forage-based systems, a second batch of calves must be purchased before any of the first batch is sold. This gives a high peak after fourteen to fifteen months. With cereal beef, first sales are at ten to eleven months, releasing money to purchase the following batch.

The slope and peak of the capital profiles will vary according to the relative costs of calves and compound.

If calves are relatively expensive compared with compound—a situation which may arise in the future—then cereal beef is likely to have a lower peak and lower average working capital than the forage-based systems. If compound is expensive compared with calf prices, then the silage and root systems will require less working capital and reach a lower peak.

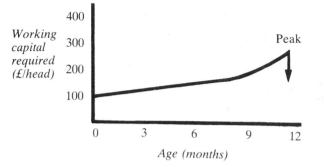

Figure 9.1(a)   Cereal beef—Capital profile

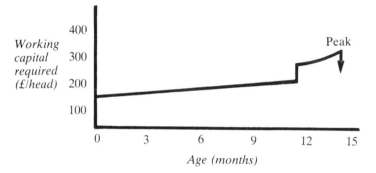

Figure 9.1(b)   Silage or root beef—Capital profile

Peak working capital for the forage-based systems may be significantly reduced by the purchase of reared three month old calves. In this situation, peak capital requirement will be reached comfortably within twelve months of purchase.

The very high stocking rates of the silage and root systems mean that both peak and average working capital required per hectare can be very large compared with the capital requirement of the grazing systems with their lower stocking rates. Even when three months old calves are purchased, the peak capital required on the grass silage beef system has been calculated by MLC to be 30 per cent higher than that of the eighteen months beef system. The average working capital needed per hectare of forage may be as much as 75 per cent higher.

In preparing budgets for all indoor beef systems, the capital requirement assumes paramount importance and interest charges need to be carefully calculated.

Chapter 10

RESOURCE COMMITMENTS

THE AVAILABILITY of land, labour and buildings are major factors determining the system of beef production to be chosen on the farm. Choose the system which most closely matches the resources available.

Land Use

Where land is in short supply, the cereal beef system will naturally be chosen since it allows all the feed to be purchased. If good quality grassland is available it may be well utilised by an efficient grazing system. Alternatively where the topography, climate and the preferences and skills of the farmer do not preclude large scale silage making, then grass silage or maize silage beef production may be the choice.

Four indoor systems known to utilise land efficiently were compared in Table 1.1 on page 11 for their potential to produce carcass meat from a hectare of land.

Figures in Table 1.1 show that compared with eighteen months beef the advantage is 31 per cent for cereal beef and 66 per cent for the silage beef system. An added advantage of silage beef may be the availability of autumn grazing after the last ensilage cut has been taken.

Building Requirements

The design of beef buildings has been covered in Chapter 3. It is however interesting to compare the housing and silo requirements of intensive systems with those of eighteen months beef. Such a comparison shows that the commonly held belief that indoor systems require more housing per head than semi-intensive systems is erroneous. This is because the throughput of housed cattle is more rapid and they utilise the houses all the year round if a batch buying system is adopted, whereas with grazing systems houses lie empty for half the year.

Housing

In order to produce a hundred finished beasts annually on the eighteen months beef system, housing is required each winter for a hundred rearers and for a hundred finishers. Cereal beef production is on an annual cycle and here the need is to house fifty rearers and fifty finishers per year. Grass silage beef results in sales at eleven to fourteen months of age. Rosemaund records point to the need for 70 per cent extra floor space in the finishing house compared with cereal beef production to accommodate those animals which are on the farm longer than twelve months. The annual requirements are shown in Table 10.1.

Table 10.1   Annual housing requirements of three beef systems

	18 months beef	Cereal beef	Grass silage beef
Accommodation in rearing house (beasts)	100	50	50
Accommodation in finishing house (beasts)	100	50	85

The floor space requirements in Table 10.2 are based on 3.0 m² per animal in the rearing phase and 4.2 m² for cereal beef and 5.0 m² for silage fed beasts in the finishing period.

Table 10.2   Building requirements for three systems of beef production to finish 100 beasts per annum

	18 months beef	Cereal beef	Grass silage beef
Cattle housing (m^2)	800	360	575
Silo space (m^2)	750	—	750

Sources: BS 5502 and Rosemaund EHF.

Assuming the same number of cattle are finished annually on each system, grass silage beef production has a lower requirement for housing than eighteen months beef.

If these two systems are compared on an equal land usage basis (say both utilise 100 ha) grass silage beef (and other forage-using indoor systems) will have the higher housing requirement because of its inherently higher stocking rate. The cereal beef system requires less buildings than the forage-based systems because of its quick throughput of stock and because the compound feed used can be purchased as required and does not have the storage needs associated with silage.

Silo space

Beasts on each of the two forage using systems require from 3.5 to 6.5 t of silage per head over their lifetime. The silage space needed is 1.5 m³ per tonne and from 5 to 10 m³ per beast.

Since the stocking rate on the grass silage beef system is double that on the eighteen months beef system the silage requirement is also doubled. This does not necessarily mean that silo capacity requirement is doubled as some producers have filled a silo with first cut grass, fed the resultant silage in June and July and refilled the silo with grass cut late in the season. This double use of silo space can reduce the high silo cost implicit in silage beef production.

Straw Usage

The bedding needed to keep cattle clean, comfortable and healthy is a significant cost in housed beef units.

Records at Rosemaund EHF indicate that on the silage beef system straw usage in fully bedded houses is approximately one tonne per head over a twelve months lifetime and 1.25 t when cattle are finished at fourteen months. In eighteen months beef units the straw requirement is approximately 0.9 t per beast finished although part-bedded yards may reduce this by about one third. Slatted floored houses eliminate the requirement for straw in the later stages of finishing.

In cereal beef units the dry diet reduces the bedding requirement to around 700 kg/head. However, animals on this system have a need for fibre in the diet and, if this is not incorporated in the compound feed, they will eat approximately one to 1.5 kg of bedding straw each day. This will increase their straw requirement to around one tonne per head.

On all-grass farms operating intensive beef systems the cost of straw is considerable and must be included when budgets are drawn up. On mixed farms where the quantity of home grown straw is sufficient to meet the needs of the livestock units the costs are restricted to those of baling and hauling.

Farmyard Manure Production

The production of faeces and urine by rearing cattle to six months old is estimated at around ten litres per day. From six months to sale (say at fourteen months) ADAS figures indicate a range of nineteen to twenty-eight litres per day. British Standards quote twenty-one litres per day in the finishing period. These figures are corroborated by weighings of manure out of the Rosemaund cattle yards and have been used in Table 10.3.

Table 10.3   FYM production per beast in the intensive beef unit

Age	Approx. volume of waste Faeces + urine (l/day)	Approx. weight of waste (t)
0–6 months	10	1.8
6–14 months	21	4.5
Total waste		6.3
Straw		1.0
Total FYM		7.3

FYM = farmyard manure
Sources: MAFF booklet 2081: Profitable utilisation of livestock manures. British standards 5502 Section 2.2 1981.
NB: 1. In this table kilogrammes have been equated with litres.
2. The figures are appropriate to cattle receiving silage as the main feed. Where such cattle are finished in the winter the waste output would be higher. Animals on the cereal beef system would produce approximately 20 per cent less manure.

Farmyard Manure Value

In Tables 10.4 and 10.5 the nutrient value of farmyard manure (FYM) is shown per tonne and per beast. No attempt has been made to value the organic matter in the manure and it may be therefore that the figures underestimate its true value. The nitrogen value used (N) is the *available* N in farm yard manure, which is taken to be 25 per cent of total N.

Table 10.4   The manurial value of FYM

	FYM Content (kg/t)	Value (p/t**)
N *(available)	1.5	58
P_2O_5	3.0	101
K_2O	7.0	124
Total		283

Source: MAFF.
NB: * Total = 6.0 kg/t.
** The cost /kg of nutrients in straight fertilisers is taken (Nov 85) as N = 38.5 p, P_2O_5 = 33.7 p, K_2O = 17.7 p.

It is true to say that many beef units, through their production of FYM, allow a saving in fertiliser costs on the farm which may in part defray bedding costs.

Grazing cattle also manure the land. However because of the very uneven spread of dung and urine the nitrogen losses from the system (both gaseous and leaching of nitrate—N) can be considerable. The beneficial effects of nutrients returned to grazing land are in any case in part balanced by the sward damage caused by treading and fouling.

Table 10.5 Approximate annual output and value of manurial plant nutrients/beast (1986)

	Output (kg)	*Value (£)
Available N	11	4.24
P_2O_5	22	7.41
K_2O	51	9.03
Total		20.68

* The above nutrient values are only fully realised if first, there are no significant storage losses arising from the leaching of nutrients. Secondly the manure is applied to the land at times of the year when crops and grass can respond to its nutrients.

Labour

Labour requirement in the beef unit is at its highest during the calf rearing stage. The labour requirement for rearing a calf to three months is estimated by Nix to be 1.9 man hours per calf per month where the best equipment is available. This gives a requirement of 5.7 hours per calf reared or 570 hours for the rearing of 100 calves. The whole time of one rearer (or half the time of two rearers) is therefore taken up while calves are on liquid feed on this size of unit. However the labour requirement is reduced where milk substitute is fed only once daily, and may be further reduced by the automated ad lib feeding of milk replacer.

Many operators choose to buy weaned calves, thereby obviating the need for this peak labour requirement. By doing this they also avoid the period in the animals' life which is most demanding of high quality stockmanship.

After weaning the labour requirement is relatively small, amounting in many units to one stockman, full or part time, helped as necessary by a number of 'floating' workers. The main unaided duties of the stockman are normally:

- Supervision of the unit including regular inspection of the stock.
- Feeding.
- Bedding-down (with help if bulls are kept).

In maize or grass silage beef units it is common for one stockman to look after 500 cattle from weaning to sale, and there are instances where this figure is 600 to 700. In cereal beef units automation of feeding is perhaps easier and the stockman may be responsible for 600 beasts with numbers close to 1000 not unknown. In the best organised units he may spend much of his time as a supervisor closely observing the health and progress of the stock.

The stockman is likely to require the assistance of other members of the farm staff on an occasional basis for the following tasks:

1. Bedding-down where bulls are kept. Two men must be present for safety reasons.
2. Handling of stock including weighing, selection and despatch for slaughter.
3. Veterinary treatments.
4. Removal of FYM.

It is impossible to be specific about labour requirement since this is affected by the beef system operated (cereal beef units commonly use the least labour per beast), the type and lay-out of the buildings, and the extent of the automation of feeding and cleaning out operations. A well designed lay-out of buildings reduces the travelling time involved in feeding and facilitates its automation. Slatted-floor housing has many critics, but it does eliminate the bedding chore and may speed the cleaning out operation compared with solid manure. At Rosemaund EHF the labour usage per beast in slatted-floor housing was one half that required by cattle in fully bedded yards.

The extra labour requirement of housed cattle
Experience has shown that in terms of supervision, feeding, bedding and all handling operations the labour requirement of the well equipped housed beef unit is similar to that of grazing beef units finishing the same number of cattle per annum. Extra labour is needed for the feeding and bedding of beasts kept indoors, and for additional ensilage operations. This is balanced by the travelling time involved in the supervision of stock at grass, and labour needed for the upkeep of fencing, for stock movements, and for the weighing and veterinary treatment of grazing cattle.

The more frequent cleaning out of cattle yards is the principal additional task in intensive units. This is however commonly accomplished at slack times of the year when labour is readily available. There is no evidence that operators of intensive beef units find the rather higher labour usage to be a serious disadvantage. Several have commented that with good management and facilities a sensible non-onerous work routine is quickly established.

Stockmanship
The systems described in this book are high input/high output enterprises and success depends upon the attainment of high daily live-

weight gains. To achieve this high performance and keep the stock healthy good stockmanship is of paramount importance. Although blue-prints can be written to point the way to technical efficiency success depends on the stockman's perceptive observation of the cattle and his 'feel' for their welfare and progress.

Changing from Eighteen Months Beef to Grass Silage Beef

It is interesting to consider the implications of the above change in terms of effects on gross margin, working capital requirements, cash flow, cattle housing, silo space requirements and labour usage. More detailed financial assessments are given in Chapter 9. The first seven of the following comparisons are taken from the *MLC Beef Year Book December 1984*, and show the changes resulting from the switch in systems on a forty hectare unit basis.

1. Gross margin/head is reduced by 18 per cent.

2. Stocking rate is increased from 3.0 to 6.0 cattle/ha.

3. Gross margin/ha is increased by 63 per cent.

4. Peak working capital/ha is increased by 27 per cent.

5. Average working capital/ha is increased by 76 per cent.

6. Cattle housing floor space requirement/ha is increased by 19 per cent (one batch/annum) or reduced by 4 per cent (continuous production).

7. Silo space requirement can be doubled/ha. However the grass silage beef system gives the opportunity for double usage of silos within the season, thus reducing the space needed.

8. Cash flow is normally more even.

9. Machinery and labour usage are higher because of the additional silage requirements and the need to clean out the houses throughout the year.

As with all system changes, careful budgeting should precede any decision, with full weight given to likely changes in working capital employed and fixed costs.

Performance Targets

Before embarking on any system of beef production the operator should define performance targets. The performance of the unit should be continuously monitored to identify any deficiencies which may arise. Remedial action can then be taken to maintain the performance in order to achieve the set targets.

The following factors are important and should be regularly compared with the system targets:

- Daily liveweight gain. An accurate cattle weigh-scale is an essential piece of equipment. It is recommended that sample pens of each age group are weighed monthly with fortnightly weighing during the sale period. Handling at these times gives an estimate of fitness.

- Feed intakes. These are difficult to measure. Occasionally the amount of feed offered to a pen of each age group should be weighed. In large beef units weigh cells may be fitted to a trailer or forage box.

- Feed quality. Batches of forages and compound feeds should be chemically analysed.

- Stocking rates.

- Mortality.

- Weight at sale.

- Age at sale.

- Carcass quality. In terms of carcass weight, killing-out percentage and EEC external fat cover and conformation classifications.

Fitting the System to the Farm

Housed systems are most attractive where:

- Grassland is prone to poaching under heavy grazing.

- Grassland is remote from the buildings or has little natural shelter.

- Grassland borders urban areas or motorways.

- Standards of fencing or water provision in the fields are poor.

- The farm is well equipped with beef houses and silos.

- Home grown straw is available for bedding.

In choosing between the proven intensive systems perhaps cereal beef should be considered first because of its simplicity. However where the objectives are to make profitable use of forage crops for beef production and to provide a cereal break the grass silage beef system is widely applicable. The use of forage maize may also be attractive but is limited geographically by soil and climatic considerations. To a lesser extent this is true also of root-based systems.

Chapter 11

MARKETING

EVEN WITH the best beef systems profits can be made or marred by inappropriate marketing. The systems described in this book make the best use of available resources. However, with increasing emphasis being placed on food quality, it is vitally important that the producer does not focus his attention solely on the physical aspects of production such as liveweight gain and Food Conversion Efficiency. Unfortunately, too little attention is paid to market requirements and to the factors which influence the value of beef cattle.

Meeting market requirements
Much of this section has been derived from a paper by Dr Monica Winstanley, Meat Research Institute (MRI).

Current demand is towards leaner beef and this move has been hastened by reports on cardio-vascular disease and its possible link with dietary intake of animal fat. A survey of consumer preferences monitored at the Royal Smithfield Show in 1982 showed a swing of 40 per cent towards leaner cuts of beef (compared to 1955). The traditional view on the need for fat to ensure good beef quality has not been substantiated by taste panels at the MRI. Current consumer preference is undoubtedly for leanness, juiciness and tenderness, all of which are satisfied by young bulls produced on the four indoor systems described in this book. There is little evidence to suggest that differences in eating quality exist between these systems.

A recent survey conducted by MLC looked at beef from two contrasting sources.

• Traditional quality beef—Aberdeen Angus heifer carcasses.

• Poor conformation beef from dairy bred Friesian/Holstein steers.

Reaction from 375 families showed a slight preference (over 8 per cent to 9 per cent) for the traditional quality beef in the categories of tenderness, juiciness, flavour and overall acceptability.

Tenderness is by far the most important attribute of eating quality, and thus meat texture dominates consumer evaluation of eating quality. Two post-slaughter treatments—initial carcass cooling and subsequent conditioning—profoundly affect texture. In practice, differences in post-slaughter handling methods override the small variations caused by breed, fatness, and finishing systems. This is particularly relevant with young bulls of low external fatness cover where toughening caused by rapid chilling (technically known as 'cold shortening') can occur.

Factors influencing the Value of Beef Cattle

These include killing out percentage (which is a function of final liveweight and carcass weight)—season of sale—fatness and conformation (carcass classification)—saleable meat yield and fat colour.

Killing out percentage is defined as the proportion of the live animal which is represented by the cold carcass.

$$\frac{\text{Dressed cold carcass weight} \times 100}{\text{Liveweight at slaughter}}$$

The dressed cold carcass weight would normally exclude the weight of kidney knob and channel fat (KKCF). If this is not deducted, then a rebate should be made of 8 kg for carcasses of up to 250 kg, gradually increasing for heavier animals.

Since killing out percentage is derived using the final liveweight of the animal it is subject to considerable variation. This can be seen in Table 11.1. The longer the fasting period between feeding and the final farm weight of the live animal the higher is the killing out percentage. This results merely from a reduction in the liveweight at slaughter, due to further loss of gut contents. The cold carcass weight remains unaffected.

Table 11.1 Variations in killing out percentage relative to period of fasting

Fasting period	Liveweight (kg)	Carcass weight (kg)	Killing out percentage
No fasting	475	250	52.6
Overnight fasting	450	250	55.6
Overnight fasting and standing in market	435	250	57.5

Thus for the same animal, killing out percentage could vary from

52.5 per cent to 57.5 per cent depending on the length of the fasting period before weighing.

The major part of liveweight loss during fasting occurs in the first twelve hours. Thereafter the weight reduction is at a slower rate. The liveweights quoted in Chapters 5 to 8 are based on overnight fasted weights (unless stated otherwise) and would equate approximately to mid-morning market sale liveweight. General guidelines based on these fasted liveweights are given for the four indoor systems for Friesian bulls in Table 11.2.

Table 11.2 Killing out percentage of Friesian bulls (based on fasted liveweights)

System:	Cereal beef	Maize silage beef	Grass silage beef Standard	Heavy	Fodder beet beef
	56	55	53	56	54

A full liveweight would give 2 per cent to 3 per cent lower values. It can normally be assumed that animals from cereal beef systems will have the highest killing out percentages because they are fed a less bulky ration and therefore contain less gut-fill.

Animals finished on the silage or root systems have a lower killing out percentage. The breed and sex of an animal will also affect the percentage. In commercial practice differences in relation to Friesian bulls are shown in Table 11.3.

Table 11.3 Effect of breed and sex on killing out percentage relative to Friesian bulls

Heifers	−1.0 to −2.0
Steers	+1.0 to +0.5
Hereford x Friesian bulls	+1.0
Charolais x Friesian bulls	+1.5
Limousin x Friesian bulls	+2.0

Although steers mature earlier than bulls they are often taken to similar slaughter weights and tend to be fatter, thus giving higher killing out percentages.

The superiority of the beef breeds, and of the Limousin cross in particular is clearly shown.

Carcass weight

Estimates of average carcass weights (excluding kidney knob and channel fat), derived mainly from information gathered on the EHFs are given in Table 11.4. The examples are for early (Hereford x

Friesian), intermediate (Friesian) and late maturing (Charolais *x* Friesian) breed types.

Table 11.4   Carcass weights on four indoor systems (kg)

| Sex | Breed | Cereal beef | Grass silage beef | | Maize silage beef | Fodder beet beef |
			Standard finish	Heavy finish		
Steer	Hereford *x* Friesian	—	215	245	230	230
	Friesian	210	225	260	245	240
	Charolais *x* Friesian	240	245	280	270	265
Bull	Hereford *x* Friesian	220	230	275	245	240
	Friesian	235	240	290	255	245
	Charolais *x* Friesian	270	255	315	275	275
Heifer	Hereford *x* Friesian	not suitable	180	—	190	190
	Friesian	200	—	—	—	—
	Charolais *x* Friesian	210	220	—	230	230

N.B. – denotes insufficient records available.

Individual carcass weights will vary by around plus or minus 10 per cent. The mean weights from each system may also vary according to the level of daily gain and the length of the feeding period.

Season of sale
Unlike grazing systems of beef production, most indoor systems do not necessarily show a seasonality of production. The periods in which the four main systems produce cattle for sale are shown in Table 11.5.

Table 11.5   Seasons of sale of finished cattle

System	Period
Cereal beef	All year round
Grass silage beef	All year round
Maize silage beef	Most commonly April–June
Fodder beet beef	April–June

The cereal and silage systems can have a distinct advantage in their ability to allow sales all the year round. This is of great benefit to the producer in terms of cash flow and yarding space (see Chapter 9) but also gives continuity of supply to the meat retailer and ultimately the housewife. Wholesaler processors and retailers (supermarkets in particular) attach great importance to this aspect.

The fodder beet system provides sales during one period only, but staff at Gleadthorpe EHF are currently looking at ways of feeding fodder beet throughout the year.

Carcass Classification
In November 1981 the EEC Beef Carcass Classification Scheme was introduced into the United Kingdom. The scheme uses five point classification scales for conformation and for external fatness. Conformation classes are described by the letters EUROP. Fat classes are numbered 1 to 5.

Since that time some of the classes have been subdivided to give the grid in Table 11.6.

Table 11.6 EEC Beef carcass classification

		Leanest			Fatness			Fattest
Conformation		1	2	3	4L	4H	5L	5H
Excellent	E							
	U+							
	−U							
	R							
	O+							
	−O							
	P+							
Very poor	−P							

Carcass classification gives four pieces of information

1. The weight of the cold carcass (in kg)
2. The sex of the animal
3. The shape or conformation of the carcass—from E (excellent) to −P (very poor)
4. The fat cover of the carcass—from 1 (lean) to 5H (very fat)

Thus the classification of a typical carcass in the United Kingdom could be, for example, 'steer—260 kg—R 4L.'

Classification helps the farmer to sell his animals at the correct level of finish. For the meat trader it acts as a basis for:

- Specifying preferred carcass type in a common language.
- Controlling variability in meat purchases.
- Pricing to encourage production of preferred carcasses.

Currently, the ideal carcass would have a fat score of three with conformation of U or better, although regional preferences do exist with regard to fatness.

Carcass fatness
As cattle become older and heavier, the feed required per unit of liveweight gain becomes greater because there is a higher maintenance requirement and a higher proportion of fat is laid down. In the final stages of finishing, the feed cost per unit of liveweight gained is often greater than the value of the additional gain (See Chapter 5). Therefore it is very important that cattle are sold as soon as they achieve sufficient finish for the market. The acceptable level of finish varies in different areas of the country, between different types of market outlet and between individual meat traders.

As an animal gets fatter its killing out percentage will tend to increase because gut size and gut fill become a smaller proportion of the total weight of the animal.

Carcass conformation
Thickness of muscling is inherent in the animal and cannot be altered by management practices. Conformation can be improved to a limited extent by a better level of finish, as fatness will pad out inadequate muscling but only at the expense of excessive trimming later on.

As the conformation of an animal improves, its killing out percentage increases marginally. This is due to a greater thickness of muscle being associated with each unit of skeleton weight and gut capacity.

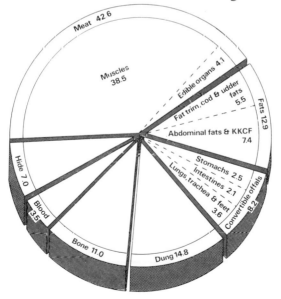

FIG 11.1   :PERCENTAGE YIELDS FROM AN AVERAGE LIVE BEEF ANIMAL

Source : M.L.C.

Saleable meat yield
Saleable meat yield is that part of the carcass which can be sold over the counter. It represents around 40 per cent of the total live weight of an animal and about 70 per cent of the carcass. This is shown in Figure 11.1.

Cutting trials by MLC have shown the average saleable meat yield for the major classification groups as detailed in Table 11.7.

Table 11.7   Saleable meat yield as a percentage of carcass weight

		Fatness					
		2	3	4L	4H	5L	Overall
Conformation	U+	74.4	73.3	72.0	70.6	68.4	71.7
	−U	74.0	72.9	71.6	70.2	68.0	71.3
	R	73.5	72.4	71.1	69.7	67.5	70.8
	O+	72.8	71.7	70.4	69.0	66.8	70.1
	−O	71.8	70.7	69.4	68.0	65.8	69.1
	Overall	73.3	72.2	70.9	69.5	67.3	

Reducing fatness from 5L to 2 has the effect of increasing saleable meat yield by 6 per cent whereas improving conformation from −0 to U+ increased yield by only 2.6 per cent.

However, improved conformation can be expected to be associated with slightly higher percentages of the more valuable cuts.

Records taken by MLC have shown that the following factors affect carcass classification:

Effect of sex
Steer and heifer carcasses most frequently fall into class R 4L, although heifers tend towards slightly poorer conformation and greater fatness. Carcasses from young bulls (in the main from cereal beef units) show conformation similar to that of steers and heifers but fatness is one class lower on average at class 3.

Effect of finishing system—Friesian bulls only
On the cereal beef system the highest proportion of carcasses fall into R3 with 13 per cent at O+ and only 1 per cent at −O. Friesian bulls recorded to date on the grass silage system show a different picture. Here, the highest proportion of carcasses has been in conformation class O+ and around 16 per cent in −O. Part of this difference may probably be explained by the generally deteriorating conformation of

black and white bulls owing to the increasing influence of the Holstein and this has been particularly noticeable in recent years. In addition lower levels of external fatness in silage beef bulls may have resulted in slightly poorer conformation. More than 80 per cent of cereal beef carcasses have graded fat class 3 or 4L, whereas only 65 per cent of Friesian silage beef carcasses achieved this level of fat cover.

Effect of breed
If taken to the same degree of finish, the different breeds and their crosses should be in the same fat classes, albeit at very different slaughter weights. In practice the later maturing breeds are not always taken to these fat levels and hence have a tendency to be leaner than the earlier maturing breeds.

Conformation classes are slightly affected by level of finish but will broadly fall into the averages shown in Table 11.8.

Table 11.8   Effect of breed on average conformation class

Breed	Average conformation class
Holstein	−O/O+
British Friesian	O+/R
Hereford x Friesian	R/−U
Continental x Friesian	−U/U+

There is little information on carcasses from different breeds on the same indoor system but some comparative figures relating to the silage beef system are available from Rosemaund EHF and from the MLC. These show that the highest proportion of Hereford x Friesian bull carcasses fall into conformation classes R and −U with fat class 3. Friesian bulls give comparative figures of O+ or R and with fat class 2 or 3.

Fat Colour

Different beef systems produce carcasses with different colours of fat. See Table 11.9.

Cereal and fodder beet beef systems provide carcasses with a very distinctive white fat colour which is considered attractive to some consumers. Grass silage beef has a cream fat colour because of the carotene in the grass. Maize silage beef external fat is an intermediate pale cream. This is often turned almost white by the feeding of a high level of compound in the final six weeks of finishing.

Table 11.9   External fat colour of carcasses from four indoor systems

System	Fat colour
Cereal beef	White
Grass silage beef	Cream
Maize silage beef	Pale cream
Fodder beet beef	White

A premium may be paid for cereal beef because it is an identifiable uniform tender product with white fat colour. The premium is highest when the general price for beef is lowest, in late summer and autumn as shown in Figure 11.2.

The premium does not compensate fully for the generally lower beef price at these times, particularly in July. If expansion of cereal beef production does occur in the future this premium, which is so vital to its financial viability, may disappear.

A premium for beef produced from grass silage is not normally paid because it is not identifiable by fat colour, but producers may be able to achieve small premia by offering wholesale meat processors a contract to supply uniform cattle all the year round.

Live versus Deadweight Marketing

The choice between live and deadweight marketing is a subject which creates considerable debate and argument amongst farmers. However, the final profitability of the beef enterprise is likely to be little affected by the decision.

The majority of beef cattle are still sold through the livestock auctions, and it can be argued that this provides the basis for dead-weight pricing. Also the producer can 'take it back home' if he is dissatisfied with the price offered. Good conformation tends to be better rewarded, and poor conformation more heavily penalised, in the live market.

Only recently have young bulls been accepted in live markets, but producers in many areas now have this outlet as an option. For steers and heifers the choice is purely one of personal preference, but it is the authors' firm opinion that, for young bulls, the outlet should be deadweight selling. The reasons for this are as follows:

• Stress prior to slaughter should be less, thus reducing the risk of the production of dark cutting meat.

• The producer can soon know whether his animals are being marketed with the correct level of finish and conformation. Armed

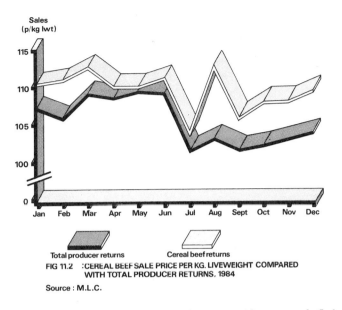

FIG 11.2 :CEREAL BEEF SALE PRICE PER KG. LIVEWEIGHT COMPARED
WITH TOTAL PRODUCER RETURNS. 1984

Source : M.L.C.

with this information, he can then rectify any deficiencies by changing his feeding management or purchasing policy.

- Safety aspects are likely to be better.
- Any premium available for barley beef animals or for the regular supply of uniform beasts is best obtained by deadweight selling.

Handling and Transport

Cattle should be handled carefully on the farm and during transport to the market or abattoir to avoid stress and injury. Young bulls in particular are prone to stress due to change of environment or mixing with strange animals, and they should not be mixed with other stock prior to slaughter.

Although loss in liveweight during transport and marketing may occur, loss of carcass weight is very unlikely in finished cattle. Experimental work has shown that very little loss of carcass weight occurs even over long journeys. It is commonly believed that the risk of weight loss is greatest with intensively fed (barley beef) cattle but there is no real evidence of this.

Summary

Beef produced by indoor methods, particularly from young bulls, should adequately meet current requirements for leanness, juiciness and tenderness.

It is important for the producer to identify reliable outlets where stock are handled correctly both before and after slaughter. In this way the consumer will be regularly supplied with a healthy, wholesome and tender product which will encourage continuing demand in the future. Satisfaction of consumer demand is the most important aspect of marketing.

APPENDIX

Hormone Implants

HORMONE IMPLANTS are still available at the time of going to press. However, in December 1985 the European Agriculture Ministers confirmed that the use of these substances as growth promoters is to be discontinued. The remainder of this section should be read with the possible ban on hormone implants in mind.

Whenever hormone implants are used it is most important to obey the rules. These are listed in this appendix under 'Cautionary Notes'.

The South West Region Beef Cattle Specialists of ADAS revised their notes on growth promoting implants for beef cattle in January 1986 and much of the following is taken from this *Technical Note No. 18*.

The rate of growth and the development of animals is affected by the balance of hormones within the body. This can be altered by administration of natural or synthetic substances and, used correctly, these can improve performance.

Among the most effective of the hormones involved are those associated with sex differences. It has now been established that the best growth rates are obtained when both androgenic (male effect) and oestrogenic (female effect) substances are present.

Females have plenty of oestrogens and so respond best to androgens. Entire males (i.e. bulls) already have androgens and so respond to oestrogens.

Castrated males (i.e. steers) do produce a little androgen and so respond to oestrogens used alone; they produce very little oestrogen and hence rarely respond to androgens unless oestrogens are given as well.

The products currently available are listed in Table App. 1.

Recommendations for bulls
Since bulls have their full quota of sex hormones and are naturally efficient converters of feed into beef it is unlikely that their response to hormone implants will be as great as that of steers.

Ralgro (Crown Chemical Company Ltd) and Compudose 365 (Elanco Products Ltd) are the only implants at present licenced for use in bulls and like all other implants these must be obtained on a veterinary prescription. In several trials Ralgro has been sequentially implanted in bulls commencing at six to twelve weeks old and at a frequency of seventy to one hundred days. It must not be implanted within seventy days of slaughter. The objective of these trials was to observe the effects of the implantations on performance, carcass conformation and behaviour.

Response in performance has varied widely. At Rosemaund EHF and in ADAS West Midlands experiments, no worthwhile response was recorded. In ADAS trials in the South West, at the West of Scotland Agricultural College and at Seale-Hayne College implants improved bull performance. In the Seale-Hayne experiments the implanted Friesian and Holstein bulls gave respectively 14 per cent and 4 per cent better liveweight gains than comparable bulls which were not implanted and in the Scottish trials the liveweight gain advantage following implantation was 5 per cent.

Carcass conformation gradings have been significantly improved by hormone implants at Seale-Hayne, at Rosemaund, at the West of Scotland College, and in trials in the United States of America.

Aggressive behaviour and sexual activity in bulls were reduced at the West of Scotland College, at Seale-Hayne and at Rosemaund but some temperamental problems still occurred. It is therefore likely that, due to their fractious behaviour and poor conformation, Friesian and Holstein bulls will show a greater benefit from implantation than bulls of other breed types. The only consistent advantages demonstrated have been in carcass conformation and in behaviour.

Recommendations for steers
Castration deprives steers of an important source of hormones, and they are therefore more responsive than bulls to the use of implants. When correctly used, implants can raise the liveweight gain and FCE of steers towards that of comparable bulls, but the carcass weight will normally be lighter.

Trials carried out by the manufacturers and by ADAS and MLC suggest that implanting two to three months before slaughter will increase liveweight gain as shown in Table App. 2.

Table App. 1 Hormone implants available in January 1986

	Finaplix	Ralgro	Compudose 200 and 365	Implixa		Synovex	
				BM	BF	S	H
Type of hormone Stock for which recommended by supplier	Androgenic Alone: beef heifers for slaughter cull cows With oestrogen: steers	Oestrogenic Steers, and bulls and heifers for slaughter; at calf, store and finishing stages, and cull cows	Oestrogenic Steers and bulls (365 only)	Oestrogenic Male veal calves	Androgenic Female veal calves	Oestrogenic Steers only	Androgenic Female cattle including cull cows
Supplier	Hoechst UK Ltd from veterinary surgeons on prescription	Crown Chemical Co Ltd from veterinary surgeons on prescription	Elanco Products Ltd from veterinary surgeons on prescription	Hoechst UK Ltd from veterinary surgeons on prescription		Syntex Agribusiness from veterinary surgeons on prescription	
Form	Cartridge containing 10 doses each of 300 mg trenbolone acetate	Pellets each containing 12 mg zeranol, in sets of 3 in magazine containing 24 doses	Silicone rubber implant containing 24 or 45 mg oestradiol 17B	Plastic cartridge containing: 20 mg oestradiol + 200 mg progesterone	Plastic cartridge containing: 20 mg oestradiol + 200 mg testosterone	Plastic cartridge containing: 20 mg oestradiol benzoate + 200 mg progesterone	Plastic cartridge containing: 20 mg oestradiol benzoate + 200 mg testosterone propionate
Maximum recommended single dosage per animal	300 mg trenbolone acetate	36 mg zeranol i.e. 3 pellets	1 implant	1 implant	1 implant	1 implant	1 implant
Theoretical active period (days)	Steers 70–100 Females 65–90	70–100	Up to 200 or 365	90	90	No recommendation	
Medicines Act Minimum Withdrawal period before slaughter (days)	60	70	Nil	90	90	Nil	Nil

Source: ADAS.

Table App. 2 **The response of steers to hormone implants during finishing**

Product	DLWG (kg)
Finaplix alone	+0.07
Ralgro alone	+0.15
Compudose alone	+0.10
Synovex S alone	+0.18
Ralgro + Finaplix	+0.25
Compudose + Finaplix	+0.25

Source: ADAS and MLC.

Repeat use of Finaplix or Ralgro is permissible provided the 'withdrawal period' laid down has elapsed. The evidence for repeat implantation is less convincing than for the effect of a single implant about three months before slaughter, but it may nevertheless be worthwhile. Combined implants are probably best not repeated. At three months before slaughter use Ralgro or Compudose 200 or Synovex S alone (or Ralgro plus Finaplix or Compudose plus Finaplix). For longer periods use at twelve months before slaughter either (a) Compudose 365 alone, or (b) Ralgro repeated at 100 day intervals (trials have shown similar results over a 300 day period from these two alternatives). In either case Finaplix can be used additionally in the last three months before slaughter and this has given a worthwhile further response in most ADAS trials.

Recommendations for heifers
At three months before slaughter use Finaplix or Synovex H (or Ralgro) alone.

Cautionary notes
There is concern in the meat trade and by consumers about the effects of any hormonal residues in meat on human health. Implants used correctly can bring substantial benefits to the beef producer, but their continued availability depends on public acceptance that they pose no possible health hazards. This is best assured by obeying the manufacturers' instructions carefully and observing the following eight 'golden rules':

- Choose an implant appropriate to the sex of the animal. The appropriate implants are listed in this appendix.

- Implant only in the correct place—normally the back of the ear and check for losses (they do occur with certain implants).

- Allow the correct withdrawal (waiting) period before slaughter or re-implantation—*this is of key importance*.

- Do *not* mix inplants unless they are compatible—*seek professional advice*.

- Use the correct dose.

- Do *not* use in an animal (male or female) likely to be used for breeding.

- Remember implants are no substitute for good feeding and management.

- READ AND OBEY the instructions supplied with the product.

All implants are now available ONLY through your Veterinary Surgeon.

Non-nutritive Feed Additives

These antibiotics work in the digestive tract, and can improve the efficiency of food utilisation. Their mode of action is quite different from that of growth-promoting implants which affect the hormone balance in the body.

Mode of action
Protozoa and bacteria in the rumen break down carbohydrates into volatile fatty acids (VFAs) which are then absorbed through the rumen wall into the blood stream and subsequently metabolised for maintenance and growth.

The three most important VFAs are propionic, acetic and butyric acids. Their relative proportions depend upon the diet, but propionic acid is produced most efficiently because it involves the least loss of methane and carbon-dioxide as waste gases.

Growth-promoting feed additives modify rumen fermentation through their effects on the rumen organisms, altering the balance of VFA production in favour of propionic acid. This means that 3 to 8 per cent more energy may be available to the animal, and this beneficial effect is independent of breed and sex and operates in all beef systems. It is also claimed that additives may have a protein-sparing effect and thereby cause a reduction in the ruminal degradation of dietary protein.

Additives available in November 1985 are detailed in Table App. 3. Romensin was introduced in the United States of America in 1974 and became available in the United Kingdom in 1977. Flavomycin

Table App. 3   Available feed additives November 1985

Additive	Active ingredient	Producing organism	Marketing firm
Romensin	Monensin sodium	Streptomyces cinnamonensis	Elanco Products Ltd
Flavomycin	Bambermycin	A group of Streptomyces	Fisons PLC Pharmaceutical Division
Avotan	Avoparcin	A strain of Streptomyces candidus	Cyanamid of Great Britain Ltd

and Avotan are more recent introductions. Further additives may reach the market in the near future.

Administration

Feed additives may be obtained either as a dilute pre-mix for incorporation on the farm or as a constituent of:

• A mineral/vitamin supplement intended for home-mixing.

• A protein concentrate to be home-mixed with cereals.

• A purchased compound used as a supplement to forage or grazing.

• A complete feed.

• A feed block.

• A calf milk powder (Flavomycin and Avotan only).

• A free access mineral (Flavomycin and Avotan only).

• A rumen bolus (Romensin only).

The dosage rate of Romensin varies with the weight of the animal, whereas that of Flavomycin and Avotan is the same at all liveweights.

The correct dose must be incorporated in the daily feed. The marketing firms have all issued informative leaflets. These should be studied and the instructions in them carefully followed.

Romensin can affect palatability and should be introduced over a three to four weeks period except when given as a bolus. When treatment is *commenced* within sixty days of slaughter it should not constitute more than one half of the recommended level throughout. Introducing Romensin too quickly or overdosing can lead to a loss of appetite and reduced feed intake. Horses should not be allowed access to feed containing Romensin—it has proved fatal!

Since these additives are not absorbed through the gut into the tissues there is no withdrawal period prior to slaughter. However they should not be fed to animals in milk or to heifers after first service.

Likely responses

Most trial work has been carried out using Romensin since it has been available longer than the other additives. The results have been notable for their variability, but independent trials, carried out by ADAS and others, have shown that the use of Romensin has usually improved the Feed Conversion Efficiency (FCE) by 4 to 8 per cent and liveweight gain by a smaller percentage.

Trials carried out at the North of Scotland College of Agriculture from 1981 to 1983 involved 1,650 animals fed mainly silage with supplementary cereals and also 470 bulls on the cereal beef system. The conclusions reached were that on the predominantly silage diets each of the three additives gave financially worthwhile increases in liveweight gain. On the cereal beef system liveweight gain was not increased, but Romensin improved the FCE. This better FCE has been reported where Romensin has been used in other cereal beef trials, but Scott (South West Region ADAS) has reported in his 1985 leaflet *Growth Promoting Feed Additives for Beef Cattle* that the greatest benefits from the inclusion of Flavomycin or Avotan may be obtained from animals fed a high forage diet. In none of the experiments has the use of additives had significant effects on carcass or meat quality or on killing-out percentage.

Recent results obtained during the 1984/85 winter at the North of Scotland College of Agriculture have shown little liveweight gain response to the use of additives, and it seems that the most frequent benefit is improved FCE arising mainly from reduced feed intake. However, with the relatively low cost of additives, even the reported modest improvements in performance are likely to be financially worthwhile in most circumstances, provided instructions for their incorporation are carefully followed.

Unlike the hormone implants the use of these additives is not at present threatened but it must be recognised that this situation could change if the demand for additive-free meat expands.

Combined use of Implants and Feed Additives

Since implants and feed additives have different modes of action it might be expected that their effects would be additive. There is however recent evidence that the combined response to the use of an additive and an implant may be greater than the sum of the responses to the two. The results in Table App. 4 were obtained from a trial involving sixty-four cattle fed a silage and barley diet (from an initial weight of 375 kg) for a twenty-one weeks period. Steers were implanted with Ralgro and Finaplix and heifers were implanted twice with Ralgro.

Table App. 4   Weight gain advantage over control (kg)

	Control	Implanted
Control	0	+10.3
Avotan	+13.2	+38.2

Source: East of Scotland College of Agriculture.

The effect on FCE of the combined use of implants and Avotan in this trial was also highly beneficial. It should however be noted that the European Ministers of Agriculture have confirmed that the use of hormone implants as growth promoters is to be banned.

INDEX

FARMING PRESS BOOKS

Below is a sample of the wide range of agricultural and veterinary books published by Farming Press. For more information or a free illustrated book list please contact:

Books Department, Farming Press Ltd, Wharfedale Road, Ipswich, Suffolk IP1 4LG.

Housing the Pig
Gerry Brent
Provides full guidelines enabling the pig farmer to assess proposals for new investment in buildings or equipment. Fifty detailed layouts are appraised.

The Sow: Improving her Efficiency – 2nd edition
P. English, W. Smith and A. Maclean
The classic manual on the practical side of the breeding, feeding, management, health and welfare of the sow and litter.

Tackling Farm Waste
Kevin Grundey
An authoritative guide for livestock farmers to help them solve the problems of muck and slurry. Covers storing, handling, disposal and profitable utilisation of farmyard manure.

The Sheep Housing Handbook
Tom Bryson
A thorough account of the techniques of housing lowland and hill sheep, from financial appraisal and the building project to health and welfare.

Intensive Sheep Management – 2nd edition
Henry Fell
An instructive personal account of sheep farming based on the experience of a leading farmer and breeder.

Forage Conservation and Feeding – 4th edition
W. F. Raymond, G. Shepperson and R. W. Waltham
Brings together the latest information on crop conservation, haymaking, silage making, mowing and field treatments, grass drying and forage feeding.

Farm Building Construction
Maurice Barnes and Clive Mander
A professional guide to the basic requirements in planning and constructing new farm buildings and modifying old ones.

Profitable Beef Production – 4th edition
M. McG. Cooper and M. B. Willis
Provides a concise account of the basic principles of reproduction, growth and development, nutrition and breeding. Emphasises production systems for dairy-bred beef.

The Principles of Dairy Farming – 10th edition
Kenneth Russell, revised by Ken Slater
The standard introduction to dairy farming covering the complete range of topics including buildings, farm systems, management, dairy farm crops and feed, milking techniques and milk production, breeding, calf rearing, disease control and profitability.

Drying and Storing Combinable Crops
K. A. McLean
An excellent reference book containing practical and fundamental details of grain drying and storage processes enabling the farmer to reach precise marketing requirements.

Oilseed Rape
J. T. Ward, W. D. Basford, J. H. Hawkins and J. M. Holliday
Contains up-to-date information on all aspects of oilseed rape growth, nutrition, pest control and marketing.

Farm Machinery – 2nd edition
Brian Bell
Gives a sound introduction to a wide range of tractors and farm equipment.

Farming Press also publish three monthly magazines: *Dairy Farmer, Pig Farming* and *Arable Farming*. For a specimen copy of any of these magazines please contact Farming Press at the address above.